DIFFRACTIVE

To all of my significant others

Diffractive Technospaces
A Feminist Approach to the Mediations of Space and Representation

FEDERICA TIMETO
Academy of Fine Arts, Palermo, Italy

Routledge
Taylor & Francis Group

LONDON AND NEW YORK

First published 2015 by Ashgate Publishing

2 Park Square, Milton Park, Abingdon, Oxfordshire OX14 4RN
711 Third Avenue, New York, NY 10017

Routledge is an imprint of the Taylor & Francis Group, an informa business

First issued in paperback 2018

British Library Cataloguing in Publication Data
A catalogue record for this book is available from the British Library

The Library of Congress Cataloging-in-Publication Data has been applied for

ISBN 978-1-4724-4545-2 (hbk)
ISBN 978-1-138-54682-0 (pbk)

Contents

List of Figures   *vii*
Acknowledgements   *ix*

Introduction   1

1   Space and Representation   17

2   Reconceiving Representation   51

3   Location, Mobility, Perspectives   85

4   Diffracting Technoscience   121

Opening Conclusions: Performing Represent-Actions   157

Bibliography   *163*
Index   *189*

List of Figures

1.1 Still from *Her*, directed by Spike Jonze, 2013   23
1.2 Still from *Teknolust*, directed by Lynn Hershman Leeson, 2002   28
1.3 Still from *Her*, directed by Spike Jonze, 2013   31

2.1 Still from *Mirror Image*, directed by Danielle Schwartz,
     cinematography by Emmanuelle Mayer, 2013   58
2.2 Still from *Mirror Image*, directed by Danielle Schwartz,
     cinematography by Emmanuelle Mayer, 2013   63
3.1 Still from *Remote Sensing*, video essay by Ursula Biemann, 2001   79
3.2 Photo of *The Transborder Immigrant Tool*, by Electronic
     Disturbance Theater, 2007   102

4.1 Advertisment for the *Misoprostol Campaign*, by Women on
     Waves, 2012   109
4.2 Photo of the performance *Epidermic! DIY Cell Lab* (foreground)
     and of the installation *Cell Track: Mapping the Appropriation of
     Life Materials* by subRosa at *Soft Power. Art and Technologies
     in the Biopolitical Age* (background), curated by Maria Ptqk,
     Amarika Project at Vitoria-Gasteiz, Spain. October 30, 2009   111
4.3 Photo of the performance Healing for La Cura (AOS), by
     Francesca Fini, 2014   122
4.4 Still from *Dying for the Other*, video installation by Beatriz da
     Costa, 2012   127
4.5 Still from *Dying for the Other*, video installation by Beatriz da
     Costa, 2012   130

Acknowledgements

I thank Yodit, for being aware and respectful of my work, for all the times that she didn't interrupt me while writing, as well as for those in which she did – beneficially – burst into the room of my own. I thank Giovanni, for having accepted my confusion and fragility, containing them in his words and eyes. And my friends Massimiliano, for being the only person I always get along with, regardless of my mood, and Caren, for having encouraged me to go on, and for being present, although too far away.

I also thank Pilar, for resting on my lap almost until the end, of her life and of my writing. Malcolm, for his positive tension. Rossellina, for her unending lickings. Cerotto, for his fluffy stickiness.

And I thank Richard, for being the most accurate, reliable and punctual person with which I have ever worked.

Introduction

This book undertakes a redefinition of the relationship between space and representation, beginning with a revised understanding of both concepts according to a performative, non-representational perspective. Here, representation is not refused, but performed differently and in reciprocity with an articulation of technospaces. I argue the latter (defined here as the sociotechnical environments in which humans and machines relate and intersect) are dynamic and contingent formations whose emergence cannot be disjoined from the generativity of the mediations that traverse them, making recourse to external, preformed representations impossible since technospaces develop together with representations thanks to the *creative capacity* at their core.

Whereas the conflation of representation and spatialisation has usually been employed to suppress the dynamism of both, acknowledging the conjoined performativity of space and representation gives dynamism back to both of them. This acknowledgement also allows a politics of spatial engagement which, as I argue through several examples in the course of the book, is precluded when space is, on the contrary, considered as an already-there quantifiable dimension, devoid of any movement. Representations of technospaces cannot be made to correspond, resist or transcend the spaces they represent because they actually *do not have space outside space*: in fact, they do not exist in a separate domain waiting to be picked up when needed, but only together with the spatial performances that they enact and within which they conversely take place.

For articulating both representation and space as well as their relationship, I primarily draw on the thought of Donna Haraway, whose politics of location (1991) and methodology of diffraction (1992; 1997), taken together in their cross-references, I consider among the most fruitful concepts to *re-turn* to, particularly when space and representation are considered in combination rather than in the singular. Both location and diffraction, while being specifically analysed in Chapter 2, are key concepts and methodological tools that traverse the whole book, both in its more theoretical parts and in the case studies which take into consideration a wide range of digital media, from film and video to locative media applications and projects for the Web.

In Chapter 1, I approach and reframe the relationship between space and representation, considering one of the most persistent dualisms upon which classical representationalism has flourished – that between place and space – in the context of an analysis of Michel Foucault's (1986) notion of 'heterotopia', to which Fredric Jameson's (1991) theorisation of 'cognitive mapping' is contrasted in order to introduce a broader discussion of mediation and the interface in contemporary technospaces. These reflections are continued and expanded in Chapter 3, where I consider the 'navigational' possibilities of contemporary geomedia, starting from the opposition between location and mobility that has animated the imagination of digital space, another dualism deriving from the conceptual opposition between space and place that I instead reconcile in order to outline a different genealogy and a politics of location for locative media.

Chapters 2 and 4 are also interconnected. In Chapter 2, I explore firstly the concept of location as a more-than-spatial term and its epistemological and political significance that has become paramount in feminist theory of the recent decades, drawing on the initial formulation offered by Adrienne Rich (1986) and arriving at Haraway's (1991) politics of situated knowledges via Sandra Harding's (2004a) concept of standpoint epistemology. Focussing on the current ecological re-vision (Hughes and Lury, 2013) of these notions, I then show how these can be used to highlight the dynamism of reality in terms of its creative, active performativity. This aspect is further explored in Chapter 4 in which, after discussing the passage from a closed to an open consideration of bodies in relation to information flows, I put Haraway's aesth-ethics of diffraction and Félix Guattari's (1995) 'proto-aesthetic paradigm' (p. 101) in dialogue.

The main section of Chapter 2 develops an investigation of the meanings of diffraction (Haraway, 1992; 1997), the fundamental concept around which *Diffractive Technospaces* revolves, that not only shows how representations can and must be articulated differently, but also the mutual relation between observing practices and observed phenomena, which only permits a performative, situated reconceptualisation of both space and representation. Literally, diffraction describes the interference of waves when they encounter an obstacle, such as when light passes through a slit.

According to the laws of classical mechanics (see Barad, 2007, p. 97 ff.), when a number of particles – imagined as little balls – are emitted by a 'ball machine' and sent through a two-slit screen, they hit the detection screen opposite the source machine according to the path that they take, that is, maintaining the separate trajectories of the two slits through which they have passed, whereas waves that pass through the two openings overlap on the detection screen, creating interference among themselves. However, quantum physics has shown that when the experiment is performed using tiny particles of matter, like electrons, they behave like waves once they arrive on the detection screen; they interfere and overlap, a phenomenon which contrasts with the wave-particle duality of classical physics.

Additionally, the experiment conceived by Niels Bohr in the late 1950s (see Barad, 2007) and effectively realised only in the mid-1990s shows that when a 'which-path' device is used in the two-slit experiment in order to understand which path an electron – imagined as a particle – takes each time it passes through the two openings, whether through the upper or the lower slit, then the electrons behave as particles once they hit the detection screen, without creating interference. This demonstrates that a given apparatus interferes in the process observed; thus, given spatio-temporal boundaries cannot be assigned to objects prior to observation, and measuring apparatuses *cut* the entanglement between the observing agent and the observed object differently each time, intimately linking 'measurement and description' (Barad, 2007, p. 109) and exposing '*an essential failure of representationalism*' to describe the presumably inherent properties of the objects observed (p. 124).

The uncertainty on which Bohr's experiment is based (Barad, 2007, pp. 115–18) does not so much postulate the unknowability of reality as the ontological performativity of epistemological practices: this means that referents do not have independent essential properties prior to the practices of observation for which, and through which, they *come to* matter. Potentially, matter exists in two states at the same time. Matter, thus, is substantially performative and informational, never inert and stable. Conversely, information and representation acquire a 'mattering' performativity as they cannot be separated from what they describe. Indistinguishable quantum states interfere, but cuts are created that configure distinctions and inhibit this interference whenever observation interferes with them.

Two physics experiments, both related to imaging, have recently brought renewed attention to the performativity of representational practices in the way observed phenomena are enacted, rather than captured. In the first one (Barreto Lemos et al., 2014), two entangled pairs of photons are created by making a laser pass through two different (down side, non-linear) crystals: each pair contains one infrared photon and one 'red' photon. One of the two infrared photons passes through an object – a silicon plate with a 3 mm cat-shaped cut-out (an homage to Erwin Schrödinger's cat paradox experiment[1]) – is combined with the infrared photon of the other pair and discarded without being detected; at the same time, the red photon of the first pair, which also combines with its twin of the second pair, although having never passed through the object that it could not even illuminate because of its insufficient frequency, nonetheless forms, together with its red photon twin, the image of the object by means of an electron multiplying charge-

---

1   Karen Barad (2007, p. 275 ff.) explains the Schrödinger cat paradox experiment – in which, simply put, a cat placed inside a closed box with a radioactive source has the same probability of dying and surviving until the box is opened and the cat's state is observed – as a classical example of the entanglement of measurement and reality: 'Suppose that we attach a device to the exterior of the chamber, which after one hour automatically opens the box and records the result of what has happened inside. Now, one would presume that after one hour, the atom has already decayed or not, and consequently the cat is already dead or not, and the recording device has thus made a record of the fact of the matter. But according to Schrödinger, this presumption would be incorrect. Rather than a resolution of the entanglement, what we have is a situation of *further entanglement* in which the recording device is now in an entangled state as well, entangled with everything going on inside the chamber. Indeed, Schrödinger argues, the entanglement persists until some cognizing subject has a look at the trace left on the recording device, which upon inspection now assumes a definite value' (p. 283).

coupled device (EMCCD) camera that, incidentally, is also blind to the infrared photons. All this can happen because the red photons, due to entangled correlation with the infrared ones, also carry the information that the infrared photons have acquired from passing through the cut-out. That is, the interference of one photon reveals information from the other photon even if the latter is not detected. So, different wavelengths can be used for illumination and for detection.

Apart from having important applications in biomedical imaging, the interest of this experiment lies in the fact that an image is obtained with a light that has never touched the figured object because information has travelled between the twin photons, only one of which contains the information on the object while only the other actually touches it. This experiment of a photograph without object shows the entanglement of information and matter in representation and thus the generativity of the latter, inasmuch as a picture is formed that is not properly a mimetic 'copy' of a bumped-into object, but the result of an *informational enacting* which nonetheless results in the same object without needing an original as its reference. Here, the relation between matter and information is shown to be one of co-emergence rather than correspondence. In fact it is only in this way that information can be extracted from undetected photons, whereas representations can be obtained from photons that have not detected the objects that they nonetheless represent.

In the second experiment (Piazza et al., 2015), a pulse of laser light hits a metallic nanowire, making its particles vibrate and move in opposite directions until a surface plasmon polariton (SPP)[2] standing wave, which is known to manifest the same wave-particle duality as light and is used as a 'fingerprint' of this duality, is created by their encounter in the middle. Here, a stream of electrons is shot that, while hitting the standing wave, also causes a change in its speed which depends on the exchange of energy packets between electrons and photons, equally foregrounding the particle-nature of light.

The experiment shows that, using ultrafast transmission electron microscopy, it becomes possible to acquire an image on a charge-coupled device (CCD) detector that captures multiple dimensions of space, energy and time simultaneously: that is, an image in which it is possible to distinguish the spatial distribution and visualisation of interference fringes on the vertical axis Y and the energy distribution and visualisation of movement on the horizontal axis X. This means that, compared with traditional 2D representation which only focusses on spatiality as an inert dimension, the observation of time and movement along with space here, or what the authors call *energy-space imaging*, are made possible since the observing method itself includes a *temporal cut* as a variable of measurement. In other words, the performative nature of the observed phenomenon, that acts both as wave and as particle, can only be 'represented' when the representational method involves a temporal variable too, which can give account not only of positions but also of relations created by the dynamism of SPPs.

---

2   Infrared electromagnetic waves used in subwavelength optics.

Thus, when Haraway retrieves representation by means of the metaphor and methodology of diffraction, she is pointing to the performativity of representations not only to show the co-emergence of meaning and matter, but also to affirm that it is always possible to materially intervene in the world's becoming by interfering with existing representations. Accordingly Barad (2003; 2007), drawing on Haraway's link between situated knowledge and optics as forms of political positioning, considers diffraction an 'ethico-onto-epistemological matter' (Barad, 2007, p. 381) that challenges the absolute separability of differences, while requiring an entangled engaging with their entangled nature that is an act of responsibility for configuring and reconfiguring the boundaries of the world with which we also connect through our visual practices.

That distinctions are not inherent in the observed objects as their properties but rather concern 'matters of practices/doings/actions' (Barad, 2003, p. 802) is a serious blow for classical representationalism and the series of dichotomies of which it avails itself. Take the example of place and space as separate dimensions, that opens Chapter 1, which mainly derives from an initial bifurcation of the theorisation of space in the absolutist tradition, drawing on Isaac Newton, in which space has been seen as a measurable and inert extension, and the relational one based on Gottfried Leibniz, in which space has been considered in its dynamism as an active and connective field of forces. The prototypical study of the fate of place that Edward Casey (1997), adopting a phenomenological approach, makes in an attempt to foreground the lively quality of embodied 'placiality' (p. 288), as he calls it, shows how very often the defences of place have been grounded on a more or less explicit series of dichotomies; such defences draw and depend on this conceptual bifurcation: first of all, that between the unlimited emptiness of space, subjected to the tyranny of time – but at the same time deprived of any relation with it – and several other displacing forces (not least technological ones), and the fullness of place characterised by inhabitability and authenticity, with the corollary of considering the latter the proper place of the Subject.

If Casey (1997) devises a 'promising resource for the revalorization of place' (p. 335) in authors going from Martin Heidegger to Luce Irigaray, Foucault included, and even in quantum theory seen as a confirmation of the 'omnilocality' and 'eventfulness' of place, which is everywhere in the universe though nowhere in particular, his theorisation, while having the merit of delinking space *qua* place from abstract representation, remains nonetheless problematic. In fact, he continues to presuppose a series of binaries, like space and place but also space and time, which his lexicon further reinforces, alternatively revolving around notions like loss and retrieval, regret and await, disappearance and rediscovery in which the lost object is, needless to say, place.

It is on such assumptions that Casey bases a critique of Foucault's '*heterotopoanalysis*' (Casey, 1997, p. 297), starting from a blatant misunderstanding of the 'imprecise' use that, in his opinion, Foucault (1986) makes of the concepts of 'site' and 'countersite'. For Foucault (1986), however, as I argue, countersites like 'heterotopias' do not antithetically redouble sites, but rework the dichotomies upon which traditional representations of space rely. This is why in Chapter 1 I discuss the usefulness of Foucault's notion of 'heterotopia' for comprehending the topological behaviour of contemporary technospaces (Lury, Parisi and Terranova, 2012), highly mediated and mediating spatio-temporal environments existing because of the reciprocal movements of emergence and counteraction of the social and the technical *at the interface*. These spaces cannot be traditionally represented, not because they are too authentic to cede their lived immediacy to the power of external signs, but because they performatively generate their representations of themselves as modes of spatialisation (Guattari, 1984; 1995; Deleuze and Guattari, 1987).

Indeed, all spaces can be heterotopic if we consider the always differently actualisable possibilities that they perform in their contingent formations, as the feminist geographer Doreen Massey (2005) asserts. In fact, following Massey, we can devise in space a 'radical coevalness', that is, a coimplication with multiplicity and difference which makes space open to change and thus impossible to represent in the absolute. The everyday mediations through which technospaces work bring to the fore their heterotopic quality as well as the several forms of practical negotiation that they require. Heterotopic representations are different from mirroring surfaces in that they are more like traversable looking glasses which engage with the place that they relate to and, in turn, ask to be engaged with (Foucault, 1986).

Contrary to Foucault's theory of heterotopias as spaces that do not reflect but displace the relations that they seem to represent, Fredric Jameson's (1991) theory of 'cognitive mapping' is a typical example of the search for another 'practice of signs' (p. 123) that, although invoked to signify the contemporary downfall of clear boundaries, nonetheless remains anchored to a representational imaginary and vocabulary. In fact, Jameson still looks for a means to re-orient oneself in the void left by the progressive distancing of society and technology – where he sees technology as the alienating other of social praxis – as well as by the collapse of the environment onto bodies that seem to have lost a clear position in the scenario (namely, the position of critical distance) and to suffocate in the impossibility of separating interiority and exteriority.

As I show in Chapter 1, Jameson's project appears contradictory because, although he hopes for a new 'situational' cartography (1991, p. 51) to oppose to the loss of anchoring points in the present, in fact he tries to rebuild it on the ashes of a Subject whom, in principle, he grants the privilege of picturing, thanks to cognitive ability, a totalising representation from a distant and still position. What Jameson fails to understand, however, is that such a safe place from which representation is produced must be included in the picture as well, given that, since the current representation of our reality mostly encompasses networks of movement, separating the place of the observing subject from the observation of these phenomena is but an a-posteriori epistemological and ideological stratagem. In other words, to elaborate a truthful picture, multiple spaces and times are purposely ignored by Jameson, which consequently justifies the plea for a representation that is also devoid of its contingency. But, as Massey (2005) again argues, every representation is always a representation of a space-time (p. 27).

In this perspective, what Jameson mostly tries to dispel, that is, the 'schizophrenic' multiplication of connections, is revealed to be what can effectively enable praxis, rather than block it in a 'claustrophobic holism' (Massey, 2005, p. 77). In fact, the multiplication of connecting nodes in contemporary technospaces enlarges the horizon of possibilities that only an engagement with the virtual capacity of an open, multiple present, in turn, allows us to devise (Guattari, 1995). But Jameson ends up transforming his ideal surveyor into a sort of explorer of technospaces with an orientalist attitude (see Gregory, 1994; Buchanan, 2005) for whom the world is still 'on display' and whose organisation still depends on forming the right, definitive picture (Mitchell, T. 1989; Bolt, 2004): someone who is able to move only along a singular trajectory (see Spivak, 1985; 1999) that does not foresee any interference along its way, not to speak of the issue of coevalness (Massey, 2005). Longing for a comprehensive representation of space, Jameson continues to keep both space and representation detached from their generative force and, thus, from history. However, linking representations exclusively with space and not with time is as problematic as defining geography as merely a spatial discipline (Zierhofer, 2002; Foucault, 2007a; November, Camacho-Hübner and Latour, 2010).

Jameson still imagines a subject who will eventually be able to cognitively and visually apprehend space as an object of scrutiny and so render a totalising picture of it. Nonetheless, neither complete transparency nor absolute opacity exists in representations and spaces, if they are not opposed but on the contrary considered as co-dependent, that is, if representations of space are not distanced from spatial practices, as some geographers such as Edward Soja (1989) and Henri Lefebvre(1991) have shown, explaining the illusion of such visual metaphors when applied to the analysis of space, which I explain in Chapter 1.

From the perspective of media theory, similar issues are initially assessed in such analyses as Jay David Bolter and Richard Grusin's (1999), who explain how both the search for an immediacy of representation and the effect of hypermediacy are but two sides of the same coin of remediation, that is, the coimplication of mediation and reality. They are echoed in the more recent discussion that Sarah Kember and Joanna Zylinska (2012) make of mediation in contemporary technospaces, where the being and becoming-with of the social and the technological can be observed. In this respect, particular importance in the field of media studies has been given to recent theorisations of the digital interface by such authors as Anna Munster (2006), Timothy Murray (2008) and Alexander Galloway (2012), in which, contrary to an instrumental view of technologies as well as to the relegation of interfacial conjunctions in an invisible seamless background, the relational and performative mediations of interfaces acting as heterotopic spaces are foregrounded.

To explain what this means, I conclude Chapter 1 with an example of an interfacial relation in which attachment and disjunction, the inside and the outside appear to be mutual and unfolding aspects: the love story between Theodore, a ghost writer of 'hand-written' letters, and Samantha, an Operating System, narrated in Spike Jonze's movie *Her* (2013). Through their evolving relationship, both the pleasures and risks of machinic connections are explored in a recursivity that eventually renders it very difficult to locate the place of the human and the place of the machine. The same logic that brings Theodore and Samantha together is also a process of differential becoming that happens contingently rather than naturally.

As Katherine Hayles (1999) clarifies in her interpretation of the human-machine relationship staged in the renowned Turing test (Turing, 2008), the latter's actuality consists not so much in the exploration of the anthropomorphic possibilities of computer intelligence as in foregrounding the cybernetic loops that make the human and the machine evolve together *and* apart at the same time, to paraphrase Barad's (2012) revealing expression, which also gives the title to the last section of Chapter 1.

Like the three female self-replicating automatons (SRAs) in Lynn Hershman Leeson's movie *Teknolust* (2002), whose interfacial nature confounds the absolute separability of the real and digital space that they ultimately cross back and forth, Samantha interfaces with Theodore in a field of distributed competencies that exceed singularity. Furthermore, Samantha is all but immaterial, not because of her feminine sexualisation, but because of her being already enmeshed with the interface that she *incorporates* rather than is inscribed onto (see Hayles, 1999; Latour, 2005). In fact, Samantha cannot be distinguished from the interface – and neither can Theodore – because their mode of existence cannot come into being independently from their ceaseless mediations.

As Theodore and Samantha's love story shows, neither matter nor information exists independently in such complex systems that, although autopoietic, need to rely on relations of alterity to evolve, as Guattari shows in his theory of schizoanalytic machines (1995), to which I return in detail in the final part of Chapter 4. In their contingent configurations, technospaces do not pre-exist their mediations, but continuously take place in an elastic, variously expandable 'middle' (Mitchell, J.W.T., 2008, p. 4), where processes of in-formation and 'mattering' combine in heterogeneous ways (see also Thrift, 2008; Grusin, 2010). Such entangled flows of distributed materiality and distributed information weaken and disenfranchise the representational imagination of spatiality that would like to keep space and representation separate as well as motionless.

This is evident, for instance, in the way maps have become so hybridised with territories 'in the middle' today that no precedence can be reasonably attributed to the one or the other anymore. In a non-representational approach, as I argue in Chapter 3, maps are not used as visual tools that figure an existing path from the outside, à la Jameson, but as 'navigational' media that can be experienced along a topological series of signs and things, information and locations taken together (Latour, 1987; Del Casino and Hanna, 2006; Kitchin, Perkins and Dodge, 2009; November, Camacho-Hübner and Latour, 2010; de Souza e Silva and Soutko, 2011; Gordon and de Souza e Silva, 2011).

An example is the series of video essays by Ursula Biemann that trace the material and symbolic routes of women at the transnational level and the way the flows of transnational economies and information and communication technologies (ICTs) intersect them, affecting their imaginary and physical mobility. Biemann's navigational cartographies are always attentive to locating the space of observation and representation 'in the field' (Biemann, 2003; 2007) – itself a highly contested term – so as to weave connections without any pretence of reducing the complexity of the observed phenomena to a totalising, undeclared perspective. Actually, they always work at articulating together the representations and the locations that they narrate, revealing a pivotal 'commitment to mobile positioning' (Haraway, 1991, p. 192; see also Parks, 2005).

Another example of navigational mapping is offered by geomedia (Thielmann, 2010) and locative media. On the one hand, their mobile, surrounding interfaces render our experience of technospaces increasingly immersive, expanding our visuality and enmeshing it with multisensorial perceptions, and in this way contribute to the reframing of representation and spatiality in performative terms (de Souza e Silva, 2006; Lemos, 2008; Kitchin and Dodge, 2011). On the other hand, they reinforce the gap between involvement and distantiation, where the latter is still assured by the maintenance of a representational approach based on the scientific, 'objective' accuracy of geospatial technologies (Gregory, 1994; Amin and Thrift, 2002; Latour, 2005; Bratton, 2013).

Interestingly, the increasing diffusion of locative mobile ICTs, having loosened most of the previous territorial 'rules' of communication (Meyrowitz, 1985), seems to prioritise location. Contrary to the always existing temptation to consider this as a return to a more delimited, binding dimension that would have been left behind by global forms of displacement 'caused' by the networks of information, however, the increasing mobile-locative aspects of contemporary media can rather highlight different ways of performing locations, assuming that not only bounded, uninformed places have never existed, but that location and mobility have always been intertwined.

The mediations of contemporary technospaces, in which ubiquitous information enhances the mobility of location – always existing but never so traceable – in unprecedented ways, gives more relevance to the 'extroverted' (Massey, 1994, p. 155), networked (Ito, 1999) and at the same time 'situational' features (Graham, 1998; Gordon, 2007; Gordon and de Souza e Silva, 2011; Massumi, 2014) of locations in formation, equally requiring that locations be localised (Latour, 2005; Meyrowitz, 2005) so as to be engaged with (Massey, 1994; 2005). This is why I conclude Chapter 3 by proposing a politics of location for locative media that does not take the locative aspects of such media for granted, but submits global positioning systems to an *actual* positioning (Parks, 2005), in which the practices and imaginaries belonging to different positionalities in technospaces are allowed to emerge. This means focussing on their material embeddedness (Hemment, 2004) and on the conflicting and uneven forces that have the power to locate as well as to *dis-locate* (Hemment,

2006) sociotechnical networks for either strategic or tactical uses, as shown in *The Transborder Immigrant Tool* (EDT, 2007–) by Ricardo Dominguez and the Electronic Disturbance Theater, a mobile phone equipped with a GPS receiver and custom software that helps Mexican migrants eschew border patrol surveillance and trespass the USA-Mexico border without being detected. The modified phone works as a performative technology that configures tactical possibilities (Zeffiro, 2015) for traversing the grid of '*hyper-geo-mapping-power*' (EDT, 2009), creating differential paths that, while refusing absolute oppositionality, work from within fixed boundaries to actualise alternative routes in hegemonic space by means of a politics of encounter and solidarity.

Notwithstanding the different declinations that the concept has assumed inside the various fields in which it has been deployed, from Science and Technology Studies (STSs) to postcolonial studies, location, upon which I linger in Chapter 2, has always been considered by feminist theory to be a historical, situated and partial dimension prioritising relationality and openness to difference and transformation. Accounting for the specificity and situatedness of material, embodied locations has been the starting point for disclosing the site of the production of knowledge, allowing for a re-vision of existing meanings and representations and unlocking the existence of alternative ones. Feminist theories of location have usually appealed to alternative representations, particularly relevant in Haraway's formulations, that not only map one's position 'in the picture', but also manifest an imaginative, creative drive towards changing the picture itself.

Despite the historical and theoretical limits listed by her critics, Rich's (1986) reclamation of the ground of politics is based on the awareness that locations are never fixed or bounded, but processual and relational both inside and outside. For Rich, if finding one's own location means dismissing the perspectives of the other, then one needs to unlearn the privilege of this space while learning that other spaces and histories also exist. Standpoint epistemology (Harding, 2004a) contributes to deconstructing such a notion of epistemic privilege, showing how different locations always involve epistemic differences. The evaluation of difference in the epistemological and political process of knowledge acquisition and the reconsideration of science in technological terms (Harding, 2008) draw standpoint epistemology very near to the position of STSs, as well as to some feminist formulations of constructivism (Lohan, 2000; Faulkner, 2001; Harding, 2008).

In particular, standpoint epistemologists are interested in the ways different regimes of truth work, as well as where and why. It is possible to give 'less false accounts' (Harding, 2004b, p. 260) of the world that are true and partial at the same time if the place where scientific knowledge *is made* is declared: in this respect, what Harding (2004c) calls 'strong objectivity' can also be thought of in terms of 'strong reflexivity' (p. 136) or 'responsible reflexivity', as Maria Lohan (2000) puts it, a reflexivity encompassing the awareness of one's own location and the way this translates into contingent practices of knowledge- and reality-making. Building on these definitions and further expanding them so as to prepare the terrain for a radical re-articulation of representation, discussed in depth in Chapter2, Haraway (1991) insists on the material-semiotic complexity of locations, and their mediated, non-innocent and non-transparent aspect made out of specific histories and critical practices. So intended, locations are surely inassimilable to either side of the classical space/place opposition, since they equally refuse the universal transparency of all-encompassing representations as well as the opacity of relativist ones.

As Christina Hughes and Celia Lury (2013) argue, the current 'ecological' re-turn to Haraway's situated knowledges does not stand as a simple repetition, but is an expansion and a turning over of previous theorisations (see also Whatmore, 2006). In what they define as 'ecological habitats', as in technospaces, the multiplication of human and non-human relations asks that we observe the social and the technical in their reversible encounters, which in turn puts into relief the productivity of boundary-making practices and 'the performativity of methodologies' as well (Hughes and Lury, 2013, p. 787). An ecological consideration of location insists on the emerging creativity of reality as 'mattering', a term which for Barad (2007) defines the ongoing performance of the world.

Barad strongly refuses representationalism, given that knowledge does not consist in the apprehension of already existing facts but in the specific configurations in which the separation between observing and observed agencies and the apparatuses of observation takes place. Boundaries are thus articulated in *cuts*, or *constraints* – as Hayles (1995) calls them in her theory of constrained constructivism – that appear as always different in relation to their spatio-temporal enactments and that render representation viable as a dynamic, engaged process rather than a static reflection, as in traditional representationalism. It is not a matter of true or false representations, but a matter of 'consistent' ones. Consistent are those representations that stand in 'diffractive' relationality, that is, in a relation of articulation, with what they represent (Haraway, 1992, p. 299).

For this reason, Barad (2007) prefers using the term 'intra-action' rather than interaction to underline the active, not merely conjunctive role of mediations in 'the mutual constitution of entangled agencies' (p. 33) that emerge in every phenomenon rather than precede it. As a matter of fact, Haraway (1992), who instead continues talking about mediations, also refers to a co-emergence of the subject and the object of representation on a shared ground of 'neverfinished ... articulatory practices' (p. 313). Actually, her proposal of an articulatory turn in representation (Haraway, 1997) is also a different theory of the mediation of both knowledge and vision, which translates into a political semiotics of representation (Haraway, 1992). According to Haraway, representation can still be employed, but only if delinked from the representational idiom (Pickering, 1994) and aligned with the performative one and with the assumptions of non-representational theory, which Haraway's theory anticipates in many respects, as I specify in Chapter 2 (Jacobs and Nash, 2003; Lorimer, 2005; Whatmore, 2006; Thrift, 2008; Anderson and Harrison, 2010).

For Haraway (1991), articulating rather than refusing representation means putting the false dichotomies that have sustained traditional representationalism in tension and, while escaping the oppositional line of reasoning, making it an epistemological and political instrument of confrontation. Haraway's articulation of representation culminates in her elaboration of a diffractive methodology that, starting from the persistence of vision in her politics of situated knowledges, passes through the notion of figuration to arrive at that of diffraction. Figuration is the term that she (1997) uses to name the possibility we have of mapping the articulations taking place at the boundaries of our realities. Drawing on Christian and Aristotelian traditions, Haraway focusses on the conjoined spatio-temporal aspects of figurations, putting into relief their strong link to location, to which they relate as constructive and transformative cartographies.

Figurations are both *topoi*, material-discursive meeting points or 'commonplaces' that can be inhabited but never in fixed or static ways, and transformative *trópoi* that continuously tend to turn, shift and displace what they figure thanks to their power of differentiation (Haraway, 2008a). Figurations are thus re-presentations that always *re-turn* as not the same (Doel, 2010; Hughes and Lury, 2013). They not only map the world as it appears, but also highlight what changes in what they map, at the same time that they change what they map and change while mapping. Their role is vital for representing the 'sociotechnical circulations' (Haraway, 1997, p. 12) in technospaces, as they manifest internal and active coimplication with them rather than supply exterior correspondences from a distance.

The most powerful figuration that Haraway (1992; 1997) uses to show the entangled performativity of reality and representation and the generative power of visual practices is diffraction, an optical phenomenon that she also employs as a methodology to interrogate the relations of light and matter, for 'mattering' light and giving light back its history (Haraway, 2000, p. 103). In Barad's (2007) words, Haraway's 'diffractive methodology is a critical practice for making a difference in the world' as well as a responsible commitment 'to understanding which differences matter, how they matter, and for whom' that puts into play the place of the observer in the observed phenomena (p. 90).

With every measurement, we look inside a phenomenon (Barad, 2007, p. 283 ff.); we do not have an outside from which to measure, so that observed differences are not so much inherent in the physical states of the observed objects but only a further extension of the entanglement, one that includes the measuring action inside the measured entanglements. This gives us the possibility of accounting for the phenomena observed in a way which enacts a contingent resolution, an 'agential separability' (p. 176) – given that no essential one exists a priori – between the observer and the observed. It follows that measuring agencies cannot measure their own entanglements with the measured object because no absolute exteriority secures objectivity, even though, at the same time, objectivity can still be performed in partial and situated accounts.

A diffractive methodology has been employed in several different fields, particularly in media studies (Lynes, 2011; 2013; Kember and Zylinska, 2012;

Smith, 2012) where it has been used to show the activity of mediations as well as the multiple power forces and perspectives at stake in the visual field. So, as an example of how a diffractive methodology can be employed to trace connections that lead to partial truths emerging from the interference of different perspectives, I conclude Chapter 2 with an analysis of the short film *Mirror Image* (2013), by the Israeli filmmaker Danielle Schwartz, whose starting point is the story of a large crystal mirror of Palestinian origin that stands in Schwartz's grandparents' house in Israel. Rather than documenting the object's story, the film actually stages the verbal confrontation of Schwartz and her grandparents around the possible origin of the mirror, presumably plundered from a Palestinian house in the course of a massive displacement of Palestinians by Israeli military forces before the birth of the Israeli State in 1948.

This mirror becomes the active interface through which a variety of both personal and collective perspectives, subjectivities and histories are diffracted through the story's characters. Usually considered a passive surface of reflection without agency, instead here the mirror acts as a trigger that unveils a chain of entangled associations (see Latour, 1994; Haraway, 2000). Over it, different points of view, which also include the narrator's internal location in the story, converge and diverge, evoking a missing picture. What is never fully represented is nonetheless paradoxically performed in a dialogic and affective process in which the characters' vision undergoes constant revision and articulation, so that truth can only emerge as the incomplete result of a declared accountability (see Butler, 1997; Barad, 2007).

The link between Chapter 2 and Chapter 4 becomes more evident as the reader proceeds towards the latter's conclusion. In fact, Chapter 4 initially centres on the dynamics of technobiopower (Haraway, 1997) and the 'constitutive practices of technoscience' (p. 35). The focus is on the multiple production and exploitation of dispersed bodies (Shields, 2006; subRosa, 2011) in the 'integrated circuit' (Haraway 1991; see also Guattari, 2000) in which networks of technobiopower run and overlap, and in which the constant modulation and exercise of information traverses bodies at multiple levels, from the biomolecular to the transnational (Deleuze, 1990; Lazzarato, 2006; Thacker, 2008; Munster, 2013). However, inside the same networks in which the flows of information are increasingly managed so that the transversal movements of communication can be kept under control, tactics of weaving connections also emerge that cut across the material-informational boundaries where hierarchies of location and identity are defined as well as fragmented.

Whereas previous forms for organising institutional and bodily spaces, such as those identified by Foucault (2003) in relation to medical knowledge, have not completely disappeared, a different imagination of autopoietic systems, based on the post-cybernetic framework of complexity (Guattari, 1995; Hayles, 1999; Parisi and Terranova, 2000; Kember, 2003; Ticineto Clough, 2003; Lazzarato, 2006; Terranova, 2006), complicates organic and mechanistic views of the body and paves the way to a non-representational and instead performative redefinition of technobodies in technospaces. If this approach does not eliminate, but on the contrary stems from the awareness of the contradictory aspects of the management of information flows with which it engages, at the same time this awareness lets the movements, turbulence and imbalances of informational environments emerge and yield alternative practices of communication.

The first example that I take comes from the 'site-u-ational' (subRosa, 2004a) approach of the subRosa collective, a group of feminist artists who work with digital and bio-technologies and privilege performance art to mimic and diffract the performances of technoscience. Staging the practices of technoscience while opening the lab to their participant public, they intend to disclose the factual production of scientific 'facts' (Latour 1987; 2004; 2010) concerning issues like reproductive technologies and cell patenting, for example, and re-present the facticity of science through recombinant art practices that at the same time demand shared involvement in their artistic actions. In so doing, they hybridise the hierarchy existing between hard science knowledges and visual arts practices, demystifying their privileged role as artists who also perform as scientists.

Working from within the same technospaces whose powers they deal with, subRosa's members create autonomous zones of resistance, which they call 'Refugia' (subRosa, 2001a), that interfere with the regulations of technobiopower and engage in responsible and situated accounts and actions. An analogous performative and participatory approach characterises the activity of Art is Open Source (AOS), a duo formed by Salvatore Iaconesi and Oriana Persico, that works on the circulation of information in urban and human ecosystems, with particular attention to the different sociospatial relations that an alternative appropriation and dissemination of knowledge can contribute to creating.

In Chapter 4, I discuss the multimedia, collaborative project *La cura*, that AOS initiated in 2012, after Iaconesi was diagnosed with a brain tumour. The project witnesses the possibility of finding an open source cure for cancer in an ecosystemic network of collaboration, which not only allows Iaconesi as patient to recover the visual and embodied complexity of his life and relations beyond the exclusory institutionalisation of his condition, but also works as a further step in AOS's reflection on the accessibility and sharing of performative – that is politically and ethically usable – information, which takes the artist's body as starting point.

The relational processes that *La cura* generates do not only regard the making of knowledge, but also the making of the subjects of knowledge that emerge in the multiple articulations of information, where life is precisely what contrasts with the closure of a system and instead reveals the unending vitality of its connections. Very similarly, Beatriz da Costa's triptych *Dying for the Other* (2012), a video installation shot in 2011 after the artist's brain surgery for a tumour which would eventually be lethal, focusses on the several 'relations of significant otherness' (Haraway, 2003, p. 8) that sustain life and that make it impossible to consider disease according to a singular, let alone human-centred perspective. At both a formal and visual level, the triptych in fact creates continuous interference between artistic and scientific practices, placing the exercises of the recovering artist's daily life side by side the exercises of medical research in the scientific laboratory which also disclose the multilayered conjunctions that bring the human and the animal, the living and the dying together in technospaces.

So, rather than being the vehicle for narrating the personal experience of disease, da Costa's and Iaconesi's bodies work like diffracting screens onto which we can see the displacement of the subject's sides and the distributed becoming of life in-between relations of alterity, which traverse them within and without. The 'politics of what', that according to Annemarie Mol (2002) is attentive to the processes and places through which information circulates and, in the case of medicine, in the ways diseases are variously enacted when they move across them, shows that locations are sites where connections are constantly made in practice, beyond the boundaries of the seeable and sayable within which the body has been traditionally confined.

Asking for 'assisted living' (FemTechNet, 2013), Iaconesi's and da Costa's works recover the body's performativity, freeing it from the constraints of a representational dimension dominant in technoscience that, severing bodies from their multiple connections, has traditionally conveyed the idea of clearly delineated organisms working properly thanks to their seemingly seamless operations of closure. On the contrary, a performative approach shows how it is not coherence but disequilibrium and continuous reconversions that link the working of bodies as assemblages to other connections that contrast the inertia of control and full coordination with the incompleteness and dynamism of a becoming-with-differences. In fact, the creativity of the mediations through which technospaces assemble can be observed only when the joints and gaps of their completed, incomplete, and yet-to-be-complete connections are made visible as a guarantee of an openness that can still be engaged with and otherwise.

A performative, non-representational approach insists on the immanent creativity of those systems whose actual form does not exhaust the virtual capacity for alternative configurations that are impossible to fully grasp according to exterior representations, because of the autopoietic, fundamentally performative, quality of a creativity that is the same in-forming force of the systems as they exist (Guattari, 1995; Lury, Parisi and Terranova, 2012). Therefore, in the final part of Chapter 4, I propose a comparison between Haraway's and Guattari's aesthetics that rests on the similar ways that they conceive creativity and imagination. Considering the aesthetics of machines, Guattari (1995) explains, does not mean either privileging the role of institutionalised arts or invoking an aesthetisation of the social, but rather gets to the bottom of the *creative nucleus* that is the same emerging force of machines when they perform, rather than express, their mode of existence. Incidentally, this is also the sense in which, in Chapter 3, I retrace an artistic genealogy for locative media (Tuters and Varnelis, 2006; Crow et al., 2008), which not so much restores the analogy between maps and aesthetic objects (Rees, 1980; Wood, 2006), but retrieves their performative, creative potential beyond their strategic and corporate applications.

Being able to grasp this incipient dimension of creation also means understanding it as a force of differentiation that works at diverting and transcending any stable representation, as in the case of Haraway's diffractive figurations (1992; 1997; 2008a; see also Stoetzler and Yuval-Davis, 2002), so as to constantly distance the identity of systems from their self-sufficiency and make encounters with alterity possible. Aesthetics, subtracted from the domain of contemplation and realigned with the processuality of the real and involvement in reality-making practices, resituates creativity inside an ecological network of practices in which the constitutive tendency toward alterity necessarily implies care and involvement with our 'companion' others, being thus redefinable as *aesth-ethics* (Haraway, 2003; 2008b). As Haraway (1991) affirms, the awareness of our partiality and of our technological, that is machinic, configuration means taking into account the creative activity where the boundaries that we build every day in shared technospaces also become the thresholds that allows us to 'trope' existing *topoi* and make differences in the world (Barad, 2007), where our search for connections is also the only possibility that we have to liberate them and make them proliferate.

Chapter 1
Space and Representation

1.1 Beyond the Binaries: Modes of Spatialisation

There has been theoretical 'efflorescence' (Agnew, 2005, p. 82) around the concepts of space and place in recent decades. The polarisation between these two concepts has permeated many debates within geography as well as other theoretical fields in both the natural and human sciences,[1] revealing many similarities in and exchanges between the metaphors used and the philosophical references employed, although for different aims. This conceptual opposition frequently underlies theorisations that treat space and place separately, as if the characteristics of space could not also be attributed to place, and vice versa, and often appears to be constructed on the basis of a more or less explicit antinomy with the excluded term, showing the utility of a separate consideration to be ambiguous, at best.

The privilege accorded to space or place in these theories seems to depend on, and in turn to generate, a series of other dichotomies that reflect the adoption of precise theoretical stances. So, for example, whereas place is usually associated with the conceptual chain of originality, authenticity, concreteness, belonging, particularity and delimitation, space is considered a more abstract concept and is aligned with a different set of ideas, including deterritorialisation, universality, boundlessness, emptiness and the absolute (Grossberg, 1996; Casey, 1997; Massey, 2005).

Generally speaking, the way space and place are commonly conceptualised today originates from two philosophical traditions: the first one, drawing on Newton, can be termed absolutist; the second one, taking Leibniz as its point of departure, is relational (Agnew, 2005, pp. 84–5). Whereas for the Newtonian tradition space is an entity in itself, independently from what occupies it and thus absolute, the Leibnizian tradition pays attention to the forces and objects that populate space and activate it.

Emblematic, in this respect, is Casey's (1997) historical survey of the destiny of place in Western philosophy, from ancient cosmogonies to the contemporary period. Casey assumes a phenomenological coimplication between place and the (embodied) self, an intimate relation that he links, for example, to Alfred North Whitehead's (1979) use of the word *region* instead of the word place to describe the body's active 'withness' in place (Casey, 1997, p. 214) and to Martin Heidegger's (1962) concept of *Werkewelt*, which denotes an enacted world of practices (Casey, 2001, p. 684). But Casey's assumptions also resonate with other

---

1 To what extent such a distinction is still valid remains an open question, and it is often around spatial issues that commonalities between them are discovered.

socio-geographical formulations, such as Lefebvre's (1991) 'trialectics' or Soja's *Thirdspace* (1989), to which we could add the several oxymoronic definitions of place in Gilles Deleuze and Félix Guattari's nomadology (1987) (for example, that of 'local absolute', cited in Casey, 1997, p. 335).

So, even though Leibniz (1956; 1973) is the philosopher who, against Descartes, reconceives extension in qualitative terms, as a concept which is no longer divisive but rather connective, Casey (1997) nonetheless criticises him for what he sees as an interchangeability of space and place in his thought – one which, while redeeming space, turns out to be 'disastrous' (p. 179) for place because place is ultimately reduced to a position, to the point of being devoid of its 'placial' quality (p. 174), which for Casey means 'the immanent scene of finite place as felt by an equally finite body' (p. 78). However, in Leibniz's philosophy, place and space cohere since space is a diffused locality and extension regards quality, not quantity; thus (and here lies his anti-Cartesianism), there are neither *partes extra partes* nor a divisive conception of extension: 'extension, when it is an attribute of space, is the diffusion or continuation of situation or locality' (Leibniz cited in Casey, 1997, p. 170). But still, Casey continues to locate a supremacy of space to the detriment of place in Leibniz's 'sea of relations', in which place *qua* position (Leibniz cited in Casey, p. 179) remains 'exterocentric to the situated subject', providing only site rather than place (p. 178). And, for Casey, place intended as *site* is a placement that remains extrinsic to what is emplaced: a point in a grid that can be calculated and represented before being occupied, as locations of cartographic visualisations are (p. 201). 'Yet, site does not situate' (p. 201), says Casey, only place does.

Deleuze (1993), in fact, is of a different opinion than Casey and does not perceive any essential contradiction in Leibniz's conceptualisation; instead, he considers these distinctions according to a theory of the modulation of forces, a differential distribution of intensities that give way to either actualisations or realisations, so that, instead of two (or more) substantial orders of monads existing and extending, there are simply different happenings of the same world. Each monad expresses the world through selection and closure but, at the same time, only in a particular place, so that each state is always a differential condition maintaining closure somewhere as an always incomplete openness elsewhere. In Leibniz, the difference between the internal and the external order is not a substantial one: 'not only is the differential relation the pure element of potentiality, but the limit is the power of the continuous as continuity is the power of these limits themselves' (Deleuze, 1994, p. 47). As Claire Colebrook (2005) also notes, saying that space is relational means that there are no absolute, pre-existing spatial conditions and that relations emerge from a process of differentiation; 'the power to differ expresses itself differently in each of its produced relations', so that the processes of spatialisation cannot be properly thought after the event, but *as* events. Consequently, 'a field is not a distribution of points so much as the striving of powers to become and that become *as this or that quality* depending upon, but never exhausted, by, their encounters' (p. 198; see also Guattari, 1995, p. 80).

Following his argument against site, however, Casey (1997) goes on to criticise the '*heterotopoanalysis*' (p. 297) that Foucault does in 'Of Other Spaces' (1986). In this brief and extremely dense text, originally given as a lecture in 1967 and later published as an article in the French journal *Architecture/Mouvement/Continuité* in 1984, Foucault discusses his seminal notion of heterotopia, against which Casey's argument is addressed. Not only does Casey find the periodisation that Foucault makes, with his distinction among the three epochs of medieval localisation, Galilean extension and contemporary situational relationality, oversimplified and inaccurate, but he also strongly disagrees with Foucault's assertion that we now live in a spatial epoch in which space specifically takes the form of 'relations among sites' (Foucault cited in Casey, p. 299). On the contrary, Casey believes that ours is definitely a dromocentric epoch, an epoch in which time as velocity dominates, where the acceleration of time is essentially seen as depending upon the technological changes of the global village. In particular, Casey's presupposition of the negativity of the term *site* makes him equate Foucault's reflection with Leibniz's 'mistake' of a 'purely positional or relational model of space or place construed as site', a concept in which space and place are also dangerously confounded (p. 299). Unfortunately, Casey laments, site as quantifiable location has become 'the dominant spatial module of the modern age' (p. 334), preparing the terrain for the subsequent hegemony of time.

The fundamental point is that, in Casey's phenomenological perspective, emplacement needs human embodiment, and vice versa. Thus, for him, a place cannot be a mere content of representation, since representation only expresses the *where* and the *what* of place but not its virtuality, which he links with the phenomenal character of place, which does not precede the inhabitation of bodies but happens only with bodies that are *oriented-to* a place that they '*might* come to' as a horizon of action, rather than a container in which they need to fit in (p. 231 ff.).

Casey's (1997) delinking of place and representation is surely noteworthy. However, notwithstanding his assumption that place belongs to the order of events rather than to the order of measurable quantities (p. 336), his defence of place against absolute space on the one hand, and the idea of a current predominance of time causing a chain of displacements on the other, still appear to be based on an ontological distinction, be it between place and space or between place and time.

It comes as no surprise, then, that Casey (1997) criticises Foucault's supposed terminological confusion among terms like 'place', 'space', 'site' and 'location' (p. 300). If we read carefully through Foucault's essay (1986), though, we can notice that, after initially defining site as a 'relation of proximity between points', he continues by saying that we cannot stop at the geometrical *whereness* of sites but need to consider the way sites are living spaces and 'what relations of propinquity, what type of storage, circulation, marking and classification' they, therefore, imply (p. 23). Foucault's claim for spatialisation is made on the basis of an analysis of the relations, or rather, the power relations, of space. In this respect, Foucault's intention is particularly evident in his final reply in an interview with the editors of the French journal *Hérodote* (2007a). After discussing the not-so-clearly outspoken role of geography in his archaeology of knowledge, he admits:

> The longer I continue, the more it seems to me that the formation of discourses and the genealogy of knowledge need to be analysed, not in terms of types of consciousness, modes of perception and forms of ideology, but in terms of tactics and strategies of power. Tactics and strategies deployed through implantations, distributions, demarcations, control of territories and organizations of domains which could well make up a sort of geopolitics where my preoccupations would link up with your methods. (p. 182)

While phenomenologists have usually centred their analyses of the heterogeneity of space on internal space, or the space of the Subject, Foucault intends to study this same heterogeneity in that which is commonly referred to as external space, the space of relations. At first glance, this appears to be a re-proposition of the dichotomy between space and place in the form of an antithesis between exteriority and interiority. But, as Foucault (1986, p. 23) very clearly asserts when writing about Bachelard's work, his intention is rather to show how the opposition inside/outside becomes an untenable one.

According to Foucault, space is by no means a void. If we stop considering space as a fixed and un-dialectical passive extension, he claims, we can also do without a conception of temporality as a phenomenon which is either intrinsic to consciousness or coincident with linear progression and organic growth. Seen in this light, then, space does not stand for the opposite of history and time, but is imbued with histories and different spatio-temporal arrangements of power relations (Foucault, 2007a, pp. 177–8) – what Massey (1994) has called *power-geometries*. 'Space itself has a history in Western experience', affirms Foucault, 'and it is not possible to disregard the fatal intersection of time with space' (1986, p. 22).

Indeed, this is particularly evident in his notion of 'other space', or space as heterotopia (1986): heterotopias, Foucault writes, are sites that 'suspect, neutralise or invert the set of relations that they happen to designate, mirror or reflect' (p. 24). Indeed, heterotopias work dynamically at relating differences among sites considered in their reciprocal counteractions, rather than as discrete entities. As will become more evident in what follows, Foucault anticipates here what Celia Lury, Luciana Parisi and Tiziana Terranova (2012) have defined as the 'becoming topological of culture', which they also call, with reference to the Leibnizian conception of space, a 'neo-monadology of trans-individuation' (p. 19). This becomes more evident, for example, in the potentialities offered by today's shifting of information into the environment, or the proliferation of ubiquitous interfaces and the various forms of digital networking and sociotechnical mediation. Actually, all these phenomena, rather than merely bringing to the fore new relations happening at the interface, underline the dynamic quality of the interfaces in being productive of such relations, and the diffuse tendency of contemporary culture to 'behave topologically' (p. 8).

Our current experience of media more and more involves a naturalisation of its technological aspects that makes it seems closer to 'real life', as devices such as transparent interfaces and touch screens demonstrate. However, the presumed seamlessness of ubiquitous technologies disguises the continuous loops of sociotechnological agents as well as the fragmentations of augmented relationality. A loop, a term borrowed from cybernetics, here indicates the simultaneous activity and recursive circulation of the technical and the social, which were previously considered as separate planes. Whereas agency has traditionally been conflated with autonomy (Suchman, 2007), the expansion and diffusion of ubiquitous technologies instead require a more supportive network of interconnected elements to work. This means that the more technologies ramify, the more their presumed autonomy – as the possibility of isolating their technicity from their context of usage and application, according to the instrumental view of technologies – is called into question (see Thrift, 2008). It follows (or rather, it simultaneously happens) that the human subject, traditionally considered as the minimum unit of the Social, loses its autonomy as the post-social condition requires 'a relational reciprocity' (Knorr Cetina, 2001, p. 530) of which individuation is a result rather than a precondition.

In a topological field, Lury, Parisi and Terranova (2012) explain, movement and relationality are prioritised over aprioristic notions of space and time, but also over society and technology considered separately. A topological behaviour is evident each time a generative spatio-temporal tendency, which is not superimposed as a model or an external force on a subsiding surface but is rather distributed along the 'threshold[s] of change' (p. 17) of a field that continuously performs its variations, is observable, as for example in the case of information over matter. Topologically considered, then, space does not pre-exist its 'set of relations', to use Foucault's expression (1986), which means that it cannot be the object of representation as long as the latter presupposes these relations as being predetermined and static.

A classic example of topology is the study of self-organised space in the manifold geometry elaborated by the mathematician Bernard Riemann in the second half of the nineteenth century as an alternative to a Euclidean measurement of space. Among others, Lury, Parisi and Terranova also refer to the notion of meta-modelisation elaborated by Guattari (1984; 1995). The latter intends meta-modelisation as an operational organisation of the world's matter as '*manner*', a 'pragmatic processuality' (1995, p. 59, emphasis added) that follows complex paths of 'virtual autopoiesis' (p. 61) which cannot be represented by recourse to external systems of signification or codification. In this sense, meta-modelisation is performative and non-representational since it disregards the distinctions between 'the semiotic machine, the referred object and the enunciative subject' (p. 30), which would work at reducing and fixing, ultimately at closing, what on the contrary is the open existential production of the plane of immanence (see Chapter 4).[2]

Analogously, Deleuze and Guattari (1987), while drawing on Riemann's theorisation for their concept of smooth space as the space that eludes metric measurement and that 'does not have a dimension higher than that which moves through it or is inscribed in it' (p. 488), insist as well that a distinction between striated space and smooth space – which should correspond to a basic space/place dichotomy, between dimension and direction, the Euclidean and the vectorial, the intensive and the extended – can only be made in the abstract. As a matter of fact, they argue, 'smooth space is constantly being translated, transversed into a striated space; striated space is constantly being reversed, returned to a smooth space' (p. 474). What cannot be properly 'placed' is, rather, their line of demarcation (p. 481).

The differences between smooth and striated space cannot be objectively identified, but depend on the way space (or place, it does not seem to matter anymore here) enfolds and unfolds (see also Munster, 2006). Thus, although, the sea is considered by Deleuze and Guattari as the example of smooth space *par excellence*, and the city appears as the always-already striated, the first one meets the necessities of dimensionality very readily, whereas the city appears to be intersected by smooth spaces, variations and transits, all over: what distinguished them is thus neither their pure extension objectively given, nor their subjective perception, 'but the mode of spatialization, the *manner* of being in space, of being for space' (Deleuze and Guattari, 1987, p. 482, emphasis added).

Later, when talking about the mathematical model of space, Deleuze and Guattari (1987) discuss 'a typology and topology of multiplicities' (p. 483) and pose a distinction between a multiplicity of magnitude, that is, a multiplicity of discrete quantities measurable according to the metrical system, and a multiplicity of distances, that is, one which is composed of a 'set of vicinities' that are reciprocally 'enveloped' in a 'process of continuous variation', so that division always implies a change in nature, and which is neither homogeneous nor dependent on metrics (p. 483). Deleuze and Guattari's distinction is, in turn, recalled by Massey (2005) who, drawing on Constantin V. Boundas's (1996) discussion of Deleuzian-Bergsonian philosophy, affirms that multiplicity can be conceived as an ensemble of discrete entities, as a 'dimension of separation', or as a 'continuum, a multiplicity of fusion' of intensities (p. 21). As Deleuze and Guattari repeat, however, such multiplicities are not antithetical, but rather envelop the one into the other. Considering them as opposed would be a failure to conceive the becoming of multiplicity as itself an engagement with the implications and negotiations of spatiality which are necessary for grounding the imaginations of space in the activity of space. As a matter of fact, a 'pure imagination' of place and space, one in which quality and quantity as well as mobility and containment are kept distinct, is, if not impossible, at least

---

2   An analogous conceptual shift, says Guattari, is that which occurs when a reference to technological machines rather than to the machinic as a meta-model is made (p. 31).

dangerous, since it excludes any possibility of negotiating with spatiality, that is, with the politics of space (Massey, 2005, p. 86).

All spaces, according to Massey (2005), contain an element of heterotopia (p. 116) because all spaces are the product of undetermined but actualisable relations, which implies that they are open and accidental *within* and *in between*. Indeed, 'it is in the happenstance juxtaposition, in the unforeseen tearing apart, in the internal interruption, in the impossibility of closure, in the finding of yourself next door to alterity', that the heterotopic quality of space resides (p. 116). For Massey, who draws on Johannes Fabian (1983), we should thus speak of a radical *coevalness* of space: space intended not as the coexistence of unity and multiplicity, but rather in co-formation with multiplicity, where locations constitute the minimum order of differentiation (which brings us back to Leibniz's assumptions). 'Coevalness', in Massey's words, 'concerns a stance of recognition and respect in situations of mutual implication. It is an imaginative space of engagement … It is a political act' (pp. 69–70). Acknowledging the coevalness of space is not only an epistemological move but also a political gesture in that, refiguring space as 'the sphere of possibility' that is co-constituted with heterogeneity and difference (p. 9), it gives space back its openness and dynamism, that is, its politics. Such politics rests on the acknowledgement of a constitutional and active difference rather than identity of space, and on the combination of a pragmatic epistemological approach that does not keep action and knowledge apart (Lury, Parisi and Terranova, 2012).

In fact, no external position guarantees the possibility of detached observation and representation of such space anymore. Rather, acting and knowing space simultaneously comes to mean *performing* space, and requires a situated approach in the context of an expanded ecological epistemology (Hughes and Lury, 2013): here, spatial simultaneity, the everywhereness of space, is never completed because *somewhere*, connections are yet to be established or unleashed. As Massey (2005) restates, 'this is a space of loose ends and missing links. *For* the future to be open, space must be open too' (p. 12).

1.2   Through the Looking Glass

Discussing the influence of the arguments of romantic philosophy on digital narratives, Richard Coyne (1999) examines the legacy of Platonic and Neoplatonic philosophy as well as the empiricist tradition in conceptualising the real in discourse on new technologies. As Coyne points out, both the theme of the unity of the real in Platonism – with all its corollaries, such as transcendence and dematerialisation – and the theme of the multiplicity of the real in empiricism equally animate debates on technology. Coyne expounds upon the encounter between romantic and empiricist perspectives by reviewing the different spatial narratives of empiricism and focussing, in particular, on their crucial relation with the issue of representation. As he argues, in the empiricist tradition, which comprises various forms of empiricism, from common-sense empiricism to pragmatic realism, space appears to be 'represented, resisted, reduced and divided' (p. 73 ff.). Specifically, when space is represented, a *correspondence* is usually sought between the representing sign and the represented object. This is, for example, the case of the use of CADCAM techniques or any other mapping technology (at least at an immediate level). But in information technology, a *resistance* to the constraints of the real is also at issue when computer representations and cyberspace are thought to offer an alternative to the limitations of the 'real world' – to its fundamental physical and spatio-temporal boundaries (see also Kember, 2003). A sharp contrast of cyberspace and real space follows from this, in which the former is imagined as the realm of boundlessness and total freedom from the constraints of the real.

As a corollary to this, (real) space can also be *reduced* to its computational representation: the premises of spatial reductionism can be fulfilled via recourse to mathematical and geometrical models taken as a point of departure. And finally, there is the fundamental question of *division*, which actually sustains the same theoretical possibility as the previous approaches that look for a correspondence, a resistance and a reduction of real space in cyberspace. It is as if recognising a substantial division between objective space as such and the subjective space of individual experience could solve the questions posed to this point, questions which are already flawed because of their need for the abstract assumption of an essential division of space:

> Space is thereby divided into the objective and the subjective, which is a distinction of some long standing. Architects and geographers commonly distinguish between space and place. A space is reducible, can be described mathematically and on drawings such as plans and maps. On the other hand, place is memory qualified and imbued with value ... The information technology characterization of this dualism affirms that there is real space to which cyberspace is set in opposition. (Coyne, 1999, pp. 78–9)

Of course, there exist other digital, as well as socio-geographical, narratives too, which Coyne (1999) discusses in detail in the second part of his book, relying on a pragmatic approach to language and on Heidegger's phenomenology (see Lussault and Stock, 2009).[3] However, the chain of binary distinctions originating in such debates turns out to be of the utmost importance in theorising both information technologies and space via the problematic linkage of *representability* (p. 106). No matter how computers are altering our conception of space and reality, whether restoring, transgressing or transcending it, the belief in representational realism haunts empiricism and techno-idealism as well.

---

3   Coyne speaks about space as involvement rather than containment – although the notion of dwelling in Heidegger is ambiguous in this respect (Lussault and Stock, 2009), and the idea of disclosure sometimes resonates with hermeneutical assumptions that still maintain the existence of a disguised reality to be unveiled.

For Coyne (1999), however, space intended as a set of lived relations requires a form of practical engagement. In this respect, he argues, cyberspace does not work by disclosing a supposed 'real reality', but by exposing the way we engage with our realities. Incidentally, even though at the moment when Coyne was writing cyberspace was still a privileged term to refer to online environments and the locative mobile ubiquity of the contemporary media environment, commonly defined as Web 2.0, makes the term sound obsolete, Coyne's arguments still remain valid for the representational imaginations of today's digital environments. Coyne uses the example of the mirror, an object normally associated with the realist attitude towards representation and often employed to describe the mirroring structure of cyberspace, showing that the mirror can also act as an interface through which the encounter with what is never out there, but which lies within, can take place (p. 226). In Foucault's words (1986), the mirror can function as a utopia, inasmuch as it is a *placeless place* which has a direct or inverted analogy with 'real place', but also as a heterotopia, since it forces the gaze to return from the virtual space behind the 'looking glass' back again to the position that the viewer occupies:

> The mirror functions as a heterotopia in this respect: it makes the place that I occupy at the moment when I look at myself in the glass at once absolutely real, connected with all the space that surrounds it, and absolutely unreal, since in order to be perceived it has to pass through this virtual point which is over there. (p. 24)

The mediation of the mirror actualises my position, connecting it to the environment that surrounds me, and at the same time shows that this recognition can happen only retrospectively, once the live event of the spatial event has given way to its representation.

Similarly Coyne (1999), illustrating the role of the mirror in Lacanian psychoanalysis and combining it with Hans-Georg Gadamer's hermeneutics (1975) and his notion of *re-presentation*, stresses the dual working of the mirror. On the one hand, it tries to capture and reproduce reality, and, on the other, it is entrapped in reflection precisely as a failure to grasp the reality 'out there', which causes an experience of splitting – the recognition of an 'otherness within' that the subject firstly experiences through its mirrored image – and distantiation.[4]

In her analysis of the false dichotomy of space and time, Massey (2005) retraces the main reason for the equation of space with stasis and closure, along with its

---

4   Without needing to fully subscribe to the view offered by Lacan (2007), which can be very problematic with regard to his theorisation of desire in negative terms, there are fruitful cues for an anti-representationalist theory of space, and of digital space, in his emphasis on the *resistance of the real*, with its dialectics of proximity and distance – which is very different from the argument of a *resistance to the real* that we find in some empiricist accounts (Coyne, 1999).

theoretical devaluation, to the association between spatiality and representation considered as a *mirror of nature*. Not only has representation been equated with spatialisation (through, for example, the written text), but space has also been accused of being anti-temporal because of its representational character. The lack of the recognition of dynamism in space passes, to cite some of Massey's examples, from the Henri Bergson of *Matter and Memory* (2004), who blames the quantitative divisibility of space,[5] through the metaphor of the blank page as that which inaugurates the 'proper' place of inscription in Michel de Certeau (1984; see also Augé, 1995, p. 85), to the political analysis of Ernesto Laclau (1990), in which space represents ideological closure, thus political non-viability, in that it precludes change.[6] So intended, though, representation is not only a fixation of space, but also of time. It is not that space is impossible to represent; it is rather that space cannot be mimetically represented (Massey, 2005, p. 28). Perhaps we should reconceive representation so as not to conflate it with immobility in the same way that we try to reconceive space as not opposed to time.

A disengagement of space from mimetic representation comes, arguably, from today's scientific practice, which reveals science as a continuous engagement toward reducing the gap between the knower and the known, that same gap which is responsible for the equation of representation and spatialisation (Massey, 2005). It is the gap that is postulated between the act of representing and the objects to be represented that, in fact, automatically generates the problem of the 'accuracy of representations' (Barad, 2003, p. 804). Wolfgang Zierhofer (2002) shows how this same gap works when geographers fall into the trap of mistaking their interpretive frames, the abstractions of what he defines as 'first order space', for the observed objects (p. 1370). Zierhofer defines first order space as the same possibility of drawing distinctions resulting from the application of a certain code: this space is composed of several dimensions that are initially un-determined and require further application – consider, he suggests, the O/I system of computer science as a prototypical example (p. 1369). Typically, we find this scheme of interpreting space to be dominant in Western Modernity, causing several dichotomies such as mind/matter or body/soul, to name but a few. Even though postmodern epistemologies seem to pluralise these rigid frameworks, they nonetheless tend to introduce, as Zierhofer notes, another dichotomy – between space and place – that

---

5 For a different interpretation of Bergson, see Deleuze and Canguilhem (2006). A residual echo of the Bergsonian dismissal of space can also be found in Kember and Zylinska (2012) when they, notwithstanding their insistence on the process of mediation, still maintain that time is what cannot be represented, and that every attempt to represent it eventually transforms it into space, proposing again the dichotomy between space and time as a dichotomy between duration and stasis, and emptying space of its intrinsic performativity.

6 Massey (2005, p. 45), in fact, asks whether Laclau's affirmation of the impossibility of closure might also imply the necessity of a different imagination of space and its relation to radical politics.

is incompatible with the non-essentialising stance according to which a space prior to observation cannot exist.

The several social, historical and cultural applications of this pre-experiential abstraction are all modalities of what Zierhofer (2002) defines as 'second order space', whose definition entails both the contingent and contextual evaluation of specific conceptions of space and the same self-reflexive possibility of distinguishing between first and second order space (p. 1370). This is why, Zierhofer argues, defining geography as a spatial discipline is not without consequences. At work here is a mistake very similar to that which traditionally relates science and truth: the presupposition that the object of epistemological apprehension exists independently from the apprehending practice. And what links the two is, again, representation. As Zierhofer asserts,

> Distinctions between matter and mind, between nature and culture, between body and soul, and between earth and heaven do not constitute a problem per se. But taken as epistemological transcendentals they are problematical, because, then, they tend to deny other possibilities, and by this they become instruments of cultural hegemony. (p. 1369)

This argument very much resonates with Hayles's definition of the 'Platonic backhand': 'The Platonic backhand works by inferring from the world's noisy multiplicity a simplified abstraction. So far so good: this is what theorizing should do. The problem comes when the move circles around to constitute the abstraction as the originary form from which the world's multiplicities derives' (Hayles cited in Massey, 2005, p. 74).

In an essay about the possibility of interpreting maps in a non-representational way, Valérie November, Eduardo Camacho-Hübner and Bruno Latour (2010) have recently affirmed that 'there is nothing especially spatial about geography' (p. 593) – at least, not when space is assumed to be an inert dimensionality.[7] Accordingly, they show how mapping practices can work at making some representations correspond to the world only when their mediations – through which both the world and its representation co-constitute each other – are kept silent. To explain what they intend, November, Camacho-Hübner and Latour also use the example of the mirror:

---

7   Which, incidentally, makes at least inaccurate, as we will see from Foucault's assertions, David Harvey's critique of Foucault's 'undialectical' conception of geography. In Harvey's (2007) opinion, Foucault, following Kant, intends geography as a spatial discipline kept distinct from history as the discipline of time. It is also what makes Foucault (2007a) pose a very subtle question to his potential detractors, noting that he does not deal with geography directly, notwithstanding the abundance of spatial metaphors in his writings: 'Can you be sure that I am borrowing these terms from geography rather than from exactly where geography itself found them?' he asks (p. 177).

> If you think about it, it is about as odd as to wonder how come there are two strikingly similar images of ourselves when we face a mirror. In effect, we have never been gazing at a world and *then* at its representation, but rather been engaging with a powerful set of intellectual technologies, so powerful that, *when viewed under a certain angle*, they project outside a virtual image of the same world with a few odd discrepancies. In other words, there exist representational techniques, and each of them produce a 'what' outside of itself that is being represented. (p. 591)

When Foucault compares heterotopias to the way mirrors work, he is affirming exactly the same thing: if, on the one hand, mirrors can be taken as surfaces that return the reality out there, considered as the source of the mimetic representations that they create through reflection, mirrors can also function as heterotopias inasmuch as they highlight the position from which, at once, representation and reality are created. Moreover, whereas utopias have an inverted correspondence to real sites that is still based on analogy, heterotopias do not need to become unreal to work differently from real space: actually, they remain located in reality, although 'outside of all places' (Foucault, 1986, p. 24).

Indeed, heterotopias disturb the syntax that holds words and things together (Foucault, 1994, p. xviii). Paraphrasing what Foucault (1983) says about René Magritte's pipe, we can say that heterotopias work at dissociating similitude from resemblance. Whereas resemblance makes representation correspond to a referential model outside, 'similitude circulates the simulacrum as an indefinite and reversible relation of the similar to the similar', escaping the 'simultaneously real and ideal monarchy' (pp. 44–5). Some examples of the disturbance of analogy that heterotopic mirrors operate can be found in Magritte's paintings *La Reproduction Interdite* (1937), in which we see a man from the back whose reflected image is in fact his back once again, or *Les Liaisons Dangereuses* (1936), in which we see a female nude partially screened by a mirror that, in fact, reflects her missing body part, but from behind. Reading them through what Foucault himself writes about Magritte, we can say that 'through all these scenes glide similitudes that no reference point can situate: translations with neither point of departure nor support' (p. 54).

Like in the case of the reflexive operation of play, which always implies a folding back onto itself that comments on what it does while it does it – as in language – the doubling operation of the heterotopic mirror collapses the dimensional distinctions that split reality in two and, rather, shifts attention to the very act that inaugurates such distinctions, where 'the composition of the map and the composition of the territory effectively coincide' (Massumi, 2014, p. 25) through a back-and-forth movement that reciprocally modulates them: 'A mutually inclusive zone of indiscernibility that doubles the affirmation of their difference with a included middle' (p. 24). Circulation, translation, modulation are all terms pointing to aspects seldom attributed to space when space has been represented as a quantifiable divisible extension and represented as an object distinguishable from, but nonetheless resembling, its representation. They are modes, on the contrary, of a performative space which escapes the means of traditional representationalism, of a different order of signs: pure signs that, since they 'refer to nothing outside their own enactment', as Brian Massumi (2014, p. 25) would put it are thus performative rather than expressive, letting both sides of the mirror accidentally co-occur in their mediation.

1.3   Of Other Technospaces

Has space been erased today by the hegemony of time, as Casey laments, or can we still maintain that we live in a spatial epoch, as Foucault contends? Does it still make sense to ask which coordinates, temporal or spatial, are more suitable for interpretation and practice within the networked society of new information and communication technologies?

Writing in a moment in which an 'overwhelming sense of compression' (Harvey, 1989, p. 240) seems to shrink time into a perpetual present, Jameson (1991) is situated among those who believe that synchronicity, as one of the main traits of the postmodern epoch, translates into a dominance of space over time, although his assumptions about space are surely very far from those of Foucault (see Soja, 1989, p. 16 ff.). Synchronicity, for Jameson, is what marks the end of history and, thus, of praxis as the possibility of intervening in our present time, and projecting ourselves into the past or the future as well. In this context, he considers technology as 'the *other* of our society': although it is not an independent causal force in itself, but rather depends on capitalism's modes of production, technology stands as 'the massive dystopian horizon of our collective as well as our individual praxis' (1991, p. 35).

In what Jameson (1991) calls the 'Third Machine Age', technologies pose a fundamental problem regarding aesthetic representation since they do not work as machines of production anymore, but of reproduction (pp. 36–7). This is a situation which requires 'a different *practice* of signs' altogether (p. 123, emphasis added), apart from an entirely different set of images used for representation. As a matter of fact, Jameson notes, the impossibility of delineating clear boundaries around places and spatial artefacts in order to distinguish between interiority and exteriority in space makes many interpretive frames based on the disclosure and expression of truth collapse. The distinctions between essence and appearance, authenticity and inauthenticity, the signifier and the signified, all cease to be useful for orienting oneself in the depthlessness of the present (p. 12).

Jameson, in fact, reads the spatiality of postmodernity through the category of 'schizophrenia',[8] which would mark the end of (uni)linear orientation in urban space

---

8   Incidentally, the geographer Franco Farinelli (2009) considers schizophrenia to be the condition of the perspective of the subject of Western Modernity, embodied by the doubting architect Filippo Brunelleschi who, under the Portico degli Innocenti in Florence, does not know whether to believe his eyes or his hands. His is an ambiguous condition that

(Gregory, 1994, p. 153). Similarly, Celeste Olalquiaga (cited in Gregory, 1994, p. 154 ff.) attributes to postmodern urban space the characteristic of 'psychastenia', according to Roger Caillois' (1984) definition, which she interprets as the confusion of the body's coordinates with the represented environment which surrounds it. This is a confusion in which the distinctions between signs and things cannot be easily re-established since recourse to a system of signs independent from reality is, in the end, impossible. As Jussi Parikka (2010) writes, psychasthenia signals 'the potentially confusing transformations of a body in non-Euclidean topologies that are not as predictable as stable architectures are supposed to be' (p. 99). In the context of Caillois' philosophy, psychasthenia delineates, for Parikka, a non-representational view of matter and the body in space, a 'haptic' approach based on the couplings and affects of bodies interacting with their environment.

Whereas Olalquiaga's position has been contradictorily interpreted as either suggesting the possibility of profiting from the crossing of boundaries and the heterogeneity of postmodern spaces (Gregory, 1994, p. 160), or of narrowing the spatial issue to a representational-visual problem (Hansen, 2006, pp. 126 ff.),[9] Jameson's (1991) position does not leave many doubts: he definitely sees this boundary's dissolution as the end of the space of praxis (pp. 25-7).[10]

Postmodern space is famously exemplified, for Jameson (1991), in the architecture of John Portman's Bonaventure Hotel in downtown Los Angeles. This building cannot be conceived as a volume anymore since it lacks a proper interior, incorporating the *en-plein-air promenade* inside itself by means of its elevators and escalators. It also lacks an identifiable exterior, its surface being covered all over with glass reflecting the surroundings. In Jameson's view, the impossibility of distancing in postmodern space is caused by the collapse of our spatial and cognitive coordinates. This marks the end of distance as well as of 'critical distance' (p. 48), rendering the possibility of abstraction, or to use a very Jamesonian term, of 'totalization', which is the very possibility of establishing connections among various phenomena in history, extremely difficult (p. 401 ff.). Nonetheless, Jameson still believes in the possibility of looking for an 'aesthetics

---

splits the unitary world prior to linear perspective, one in which 'the map not only kills the Earth but also humiliates its language, since it rigidifies not only the object but also the way the object is referred to, paralyzing the subject as well' (Farinelli, 2003, p. 79, author's translation).

9   Reading Caillois' concept as 'a shift in the economy of embodiment and representation' (p. 131), as Parikka (2010) later does, Hansen (2006) goes on to criticise Elizabeth Grosz's position on psychasthenia as well, as she 'rather than seeing psychasthenia as an expression of the "originary" condition for dynamic bodily self-movement in the concrete form of today's technics, … can only understand it to mark a failure in the representational meshing of subject and body that, on her account, forms the condition for bodily activity' (pp. 129–30).

10   On a completely different plane, the schizoanalytic cartographies that Guattari (1995; see Chapter 4) proposes possess a creative, constructive operationality that is what effectively *allows* praxis, rather than blocking it, as Jameson believes.

of cognitive mapping' as the only way to 'hold the truth of postmodernism' (p. 54). He understands that the new representational faculty for mapping the space of postmodernity cannot be mimetic anymore (1988, p. 348; 1991, p. 51), and his appeal to mapping does not imply, in principle, recourse to the presumed truth-function of traditional maps.

Following Kevin Lynch's (1960) analysis, and somehow also echoing Lefebvre (1991), Jameson (1991) rather invokes a 'situational representation' (p. 51) that correlates 'existential data (the empirical position of the subject) with unlived, abstract conceptions of the geographical totality', in the search for a 'new intermediate space' which is liveable and generates a 'new Utopian spatial language' (p. 128). Even so, he seems to be continuously alternating between the acknowledgement of a different tension between the multiple axes of global networks, which do require a different perceptual attitude, and regret for the loss of historical continuity; the latter, he writes, has been substituted with what he sees as a 'game board' (p. 373) of isolated units, whose spatio-temporal discontinuities remain, in the end, unrelated, although he pointlessly tries to catch or compensate for them with, again, recourse to the unifying faculty of vision (see also Bhabha, 1994; Holmes, 2000; Ciccoricco, 2004).

Unfortunately, what makes Jameson go round in circles is that it is not so much postmodern space that is in motion, but rather the space occupied by the postmodern 'subject' (Buchanan, 2005, p. 19). Whereas, in most cases, the crisis of representation occurs among the sites to be represented, rarely is the site 'from which that representation emanates', in fact, taken into consideration (Duncan, J.S. 1993, p. 39). However, both sites are, literally speaking, sites of representation. It makes perfect sense, then, that Ian Buchanan (2005) compares Jameson's spatial survey to Gustave Flaubert's orientalism (see also Mitchell, T. 1989), or that Derek Gregory (1994) calls Jameson a 'surveyor within the hyperspaces of its new technoculture' (p. 161). In fact, Jameson employs all the representational tropes of modern explorers and ethnographers who, entrusting vision as the most reliable sense, engage in a vast array of fieldwork practices to detect, name and classify otherness, without really relating to it.[11]

---

11 Many critiques have been advanced against Jameson's theorisation of cognitive mapping, particularly against his universalisation of the predominantly European and North-American phenomena of postmodernism and late capitalism, and the bourgeois-humanist assumptions of his arguments (Bhabha, 1994; Massey, 1994; 2005; Carr, 2004; Mirrlees, 2005; Spivak, 1985; 1999). According to Gayatri Chakravorty Spivak (1999), for example, the universal assumptions behind Jameson's cultural description repress heterogeneity because they do not account for the position from which they are made (p. 314). Jameson, she argues, not only passes off the economic logic of microelectronic capitalism as universal, but its cultural logic, too (p. 334), while at the same time establishing an isomorphism between modes of production and cultural expressions, which, again, follows the Marxian distinction between base and superstructure. Moreover, the differences that Jameson's totalising framework takes into consideration (Spivak, 1999, p. 406) are interpreted as being merely class realities, with class being the only category of analysis

Massey's (2005) critique of Jameson's totalising vision, which resonates with that of Homi Bhabha (1994), is particularly relevant. She identifies the dominant imagination of globalisation with a 'depthless horizontality of immediate connections' that usually reflects a 'totally integrated world' (p. 76), although it must be noted that, in Jameson's view, this instantaneous simultaneity is perceived as a sign of disintegration. The priority accorded to space as immediacy in similar analyses, while not necessarily leading to a positive evaluation of the spatial category, is very often accompanied by its subordination to time (Massey, 1994; 2005), along with the entire constellation of the gendered, sexist distinctions associated with this binary and which are still operative today (Massey, 1992). Space is either reduced to a structure upon which power exerts its forces, or it is chronologised inside a timeline which privileges (and laments the loss of) Western history as the only locus of agency (Grossberg, 1996, p. 177). What is more, as Massey (2005) comments, drawing on Lawrence Grossberg (1996), history is intended as a singular, unified trajectory – a theoretical fiction, as postcolonial and subaltern studies contend (Featherstone, M. 1993; Bhabha, 1994).

By reflex, spatiality too is conflated with unity and singularity: if the space of the present is the space of immediacy, and for this reason it is very difficult to temporalise according to such an idea of linear and progressive time, this immediacy is thought to bring together static objects which already lack any dynamism *per se*. Spatial differences are reinscribed in the sequentiality of a hegemonic time as if they were stages of the same evolutionary process (Massey, 2005, p. 68) and thus left behind (or before), rather than seen in their coeval differences (see Duncan, J.S. 1993, pp. 46–7). Unfortunately, as Gupta and Ferguson perfectly synthesise, 'the presumption that spaces are autonomous has enabled the power of topography to conceal successfully the topography of power' (cited in Massey, 2005, p. 67).

But space is neither traversed by nor congealed in time; it is itself temporal and historical. As Massey (2005) puts it,

> to read interconnectivity as the instantaneity of a closed surface (the prison house of synchrony) is precisely to ignore the possibility of a multiplicity of trajectories/temporalities … a claustrophobic holism in which everything everywhere is already connected to everywhere else. And once again it leaves no opening for active politics. (p. 77)

As an alternative, Massey proposes that we read space as the sphere of radical heterogeneity where multiplicity and space are co-formed along multiple lines of power that create connections as well as disconnections (p. 99 ff.). For Massey, only the consideration of such negotiations opens up a space which is truly political

---

(p. 52; see also Bhabha, 1994, p. 219). And while some read the theorisation of cognitive mapping as a way to recentre the political subject of late capitalism (Mirrlees, 2005), others (Carr, 2004) read in Jameson's appeal to universality a regretful nostalgia for the centred subject of Western Modernity (Spivak, 1985; 1999).

in its quest for active engagement, a space *in the making*. Whereas immediacy and connectivity can evoke a unifying dimension ruled by a singular time, in which everything must converge or otherwise absolutely diverge, coevalness, on the contrary, for Massey, suggests a co-existence of different spatio-temporalities in which difference is not measured in terms of distance, as incommensurability, but in terms of contemporaneity, as confrontation (see also Chow, 2006).

It is now worth going back to Foucault (2007a) for a moment, so as to reformulate a critique of Jameson through his words. Among the main characteristics of heterotopias, Foucault actually includes temporality. Heterotopias are linked with heterochronies, which means that they change over time. That we live in a spatial epoch, thus, has a very different meaning for Foucault than it has for Jameson. As the former writes:

> For all those who confuse history with the old schemas of evolution, living continuity, organic development, the progress of consciousness or the project of existence, the use of spatial terms seems to have the air of an anti-history. If one started to talk in terms of space that meant one was hostile to time. It meant, as the fools say, that one 'denied history', that one was a 'technocrat'. They didn't understand that to trace the forms of implantation, delimitation and demarcation of objects, the modes of tabulation, the organisation of domains meant the throwing into relief of processes – historical ones, needless to say – of power. (p. 178)

On the contrary, the image of 'totalised globality' that Jameson outlines does not leave space for the processuality of space, and in so doing occludes the possibility that different spatial forces ever come to light. Space cannot be encompassed by the totalising representation that is still invoked by Jameson in that same moment that he recognises it as impossible. Such impossibility, however, does not put an end to representation *tout court*, but challenges a specific kind of representation: mimetic representation, to be precise.

As a matter of fact, contrary to what Edward Relph believes (1992), heterotopias do not *represent* postmodern space more than places represented modern space, precisely because heterotopias do not rely on mimetic representation. As we have already seen, the latter does not only work at identifying and fixing space (as well as time): it also presupposes that space stands somewhere (or, which is the same thing, everywhere) as a static object ready to be reflected in representation. Learning from heterotopology, it is perhaps time to turn to a different representational functioning, one which sees representation not as a reflecting mirror, working as a perpetually deferred utopia as in Jameson, but as a heterotopic mirror, that is, a 'looking glass', which performatively engages with the place it relates to and can, in turn, be actively engaged with. This would also resolve the issue of the loss of distance that Jameson laments: why, in fact, mourn the loss of distance as the loss of the very possibility of representing space, as Jameson does, rather than take it as an opportunity to understand the impossibility of the very existence of space as the distant object for an observing subject?

In their socio-geographical reflections on space, both Lefebvre (1991) and Soja (1989) very similarly take into consideration the two illusions governing the misconceptions about space which prevent us from thinking about its inherently transformative, relational qualities. Lefebvre (1991, p. 27 ff.) speaks of 'the illusion of transparency' and 'the illusion of opacity'. The former presupposes that mental space is distinct from social space, with human comprehension deciphering space in order to render it perfectly legible (see also de Certeau, 1984). The known coincides with the transparent, and obscurity of sense and space cannot legitimately exist. This transcendent assumption (see Coyne, 1999; Grusin, 2000; Wertheim, 2000) 'identifies knowledge, information and communication. It was on the basis of this ideology that people believed for quite a time that a revolutionary social transformation could be brought about by means of communication alone' (Lefebvre, 1991, p. 29). On the other hand, 'the illusion of opacity', which manifests a materialistic attitude, imagines space as a 'substantial reality' that resists representation until it is eventually overwhelmed. However, 'each illusion', Lefebvre continues, 'embodies and nourishes the other', so that they never exist as such, but continuously support and recall each other (p. 30).

Soja (1989), like Lefebvre, counterposes 'opaqueness' and 'transparency' as the main causes of the misrecognition of the social production and reproduction of spatiality (p. 122 ff.). An 'empiricist myopia' and a 'hypermetropic illusion' (let us note the use of visual metaphors once again) are, for him, the major causes of a persistent dualism in the theorisation of space, seen alternatively as a measurable substance, according to a 'short-sighted approach' governed by an objectivist presumption, and as an 'over-distancing vision' guided by the subjectivity of cognition (the lost critical distance of Jameson).

Analysing these same issues in relation to digital technologies and technospaces, Bolter and Grusin (1999) date the faith in an '"interfaceless" interface', which will eventually do without mediation, back to the Renaissance, in which they locate the origin of the 'aesthetic value of transparency' manifested in the metaphor of the window used to describe linear perspective, which also accompanied the initial applications of the graphical user interface (or GUI, pp. 23–4, 31). The historical counterpart of this desire for immediacy is the logic of hypermediacy that, rather than seeing representation as a window open toward the world, sees representation as itself 'windowed' (p. 34), or as a coexistence of multiple points of view. The logic of hypermediacy is not only aware of, but also extremely fascinated by mediation, with which it plays, as we can see from Medieval manuscripts, Baroque cabinets of curiosities, trompe-l'oeil paintings and the collages and photomontages of the twentieth century, among the various artistic expressions of hypermediacy in the Western tradition (p. 34).

But things are not quite so neat: actually, Bolter and Grusin (1999) point to the interdependence between immediacy and hypermediacy: 'just as hypermedia strive for immediacy, transparent digital technologies always end up being remediations, even as, indeed precisely because, they appear to deny mediation' (p. 54). Although Kember and Zylinska (2012), via Hayles, counter Bolter and Grusin, arguing that they still see media 'as fulfilling a particular socio-cultural need, and as arriving only in response to this already preformed need' (p. 9) and continuing to locate agency in the human subject alone, they nonetheless recognise that Bolter and Grusin pose the 'inseparability of mediation and reality' in their notion of remediation (p. 10): not only does remediation show that reality and mediation cannot be separately conceived or practiced, it also shows that (digital) technologies do not change the *status quo* of reality as much as they remediate its previous mediations in an uninterrupted chain.

Interestingly, Bolter and Grusin (1999) also refer to Marc Augé's (1995) definition of non-places to reinforce their argument: 'Cyberspace is not, as some assert, a parallel universe. It is not a place of escape from contemporary society, or indeed from the physical world. It is rather a nonplace, with many of the same characteristics as other highly mediated nonplaces' (p. 179). Calling cyberspace a non-place may seem at odds with their previous arguments about mediation in this regard. However, they are not subscribing to the popular view that counterposes real and virtual spaces, or places and non-places, as two distinct spheres; instead, they consider the imbrication of places and non-places according to Augé's proper definition of non-places, which the latter in turn claims to have derived from de Certeau's (1984) distinction between place and space.[12]

To put it briefly, de Certeau (1984) distinguishes, on a pragmatic and not on a morphological level (p. 126), between the geometrical and anthropological quality of spaces, seeming only apparently to rest on a dichotomy which he nonetheless turns upside down. Places, for de Certeau, comprise stable identifications of territories and subjectivities which coexist in a coherent unity (p. 117). Spaces are, instead, practised places, traversed in any direction by heterogeneous trajectories in which identity continuously passes into alterity and vice versa, breaking the consistency and simultaneity of homogenous time and space (p. 102). If places are 'determined' through 'objects', spaces are instead determined through 'operations' and, he points out, 'between these two determinations, there are passages back and forth' (p. 118). For example, even though 'tours' are generally associated with the representation of space, *acted* and going through different itineraries, and 'maps' with the representation of place, *seen* and depicted on a plane of projection (p. 119), the daily practice of space comprises, in the end, a continuous interlacing of the two. Even better, the itinerary, rather than being an alternative to the map, is, for de Certeau, the very condition of possibility from which it has progressively disengaged, disguising its relation with heterogeneity. It is worth noting that de Certeau, just like Foucault, also uses the metaphor of the mirror to explain what practising spaces means. As it

---

12   Even if a symmetrical parallel between these pairs of binaries (de Certeau's space/place and Augé's non-place/place) cannot be exactly drawn (Augé, 1995, p. 79 ff.), both theorists do not pose an ontological distinction between spatial forms as much as they pose two different senses of the representation and practice of space.

is for the child, who in front of the mirror experiences its duplicity, being at the same time the one and the other of reflection, 'to practice space is … in a place, to be other and to move toward the other' (p. 110).

This otherness of space is what renders space radically heterogeneous from the very beginning. It sounds perfectly reasonable, then, that Augé (1995) affirms that non-places are the opposite of utopias (p. 111). Differently from utopias, non-places do not welcome any organic society (Jameson's lost myth) and they also exist *in actuality*. One of the reasons why, for example, non-places such as airports, airplanes and stations are among the favourite targets of terrorist attacks, according to Augé, is precisely because they negate the ideal of a unified, perfectly enclosed territory. Like heterotopias, non-places are, indeed, heterogeneous sites comprising internal contradictions that do not 'hold' together (pp. 110–11).

1.4   Together-Apart (as One Movement)

The dichotomies deployed around the mediating interface between transparency and opacity, dynamic form and inert matter, the object and the subject of knowledge, together with the contradictions that they generate, emerge and are often even sharpened not only when society and space are at issue, but also when technology is at stake (see Kember, 2003): that is, when the dynamics of *technospaces, as those spaces where society and technology co-occur in their mediations*, are considered. It is no coincidence that the increasing significance of the practices of mediation in topological thinking is strictly related to the articulations of the social and the technological (Lury, Parisi and Terranova, 2012, p. 9) where, screens, frames, borders and interfaces are seen as dynamically remaking the relations of what they put in connection and make happen at the same time (see Grosz, 2008).

In this context, the notion of interface can be an extremely problematic one, and needs resituating. As Munster (2006) notes, the usual way this term is employed in the context of digital technologies evokes an opposition between the human and the machine, which usually presupposes an instrumental conception of technology as tool (p. 47). However, the digital interface is neither an opaque interface (usually figured as a mirror) that we bump into nor a transparent interface (usually figured as a window) that we must overcome. Rather, the idea of a disappearance of the interface can also lead to a reinforcement of anthropomorphism (p. 125), working as a counterpart of the instrumental argument. In any case, both perspectives turn away from the acknowledgement of the interpenetration, the topological enfolding of both sides of the interface, and the creation of an active field of negotiation in-between.

As Foucault (1986) shows when talking about the mediations of the mirror as heterotopic interface, it is only by going back and forth through the looking glass that the two sides of the mirror can be perceived as co-implicated before being represented as separate. Interfaces working heterotopically stand in different contexts as tangible interfaces, where they also exert 'a counteraction' – for example, the ubiquity of interfaces in smart urban environments today – and at the same time they give the mediated character of their environment back to the users, so that the former too appears as both absolutely real and indefinitely virtual (p. 24).

If it is made to work in this way, an interface is not of the order of representation, but is more similar to what Munster (2006) intends when she proposes a reconception of the digital interface in a Baroque, Leibnizian and topological framework rather than a classical one, that is, not according to modes of separation from materiality but according to modes of interdependency and pliability, or what she calls 'a folded sense of the digital' (p. 47). Analogously, Murray (2008) highlights 'the role played by the new screen arts in addressing the paradigm shift away from the remnants of humanist visions of subjectivity and projection' (p. 9) toward what he terms 'the digital Baroque': a move that distances the screen from the representational idea of a surface of projection and instead looks at the intra-activities that take place through the 'machinic' operations of the screen as a dynamic interface.

Such intra-activities are what Galloway (2012) refers to when he talks about the 'intraface', a term which evokes that of intra-action used by Barad (2007), which will be discussed at length in Chapter 2 of this book. To anticipate it very briefly, Barad develops this notion regarding the observation of phenomena related to physics, and quantum physics in particular. She considers intra-action as a process of differentiation that does not pose the related elements in a relation of 'radical exteriority, but of agential separability', an 'exteriority within' where togetherness and separability are the mutual sides of, precisely, an intra-action (2012, p. 32).

Similarly, Galloway (2012) uses the term intraface to designate an interfacial activity of mediation *within* the interface which exceeds the object-status of the interface standing in between its edges and centre, in a 'zone of indecision' (p. 40) where the inside is implicated in the outside and vice versa. This effect twists the use of the interface from a representational modality of expression (according to a source-execution trajectory) to one of performativity, which does not presuppose a point where what happens at the interface originates from, but where the social and the technical, the human and the non-human are mutually mediated back and forth (p. 42; see also Massumi, 2014).

Such couplings and splits of the human and the machine at the interface, rather than the promises or the dangers of their connections, are at the centre of *Her* (2013), a movie written and directed by Jonze that focusses on a love story which, incidentally, is also a love story between a man and an Operating System. The two protagonists, Theodore, a flesh and blood 39-year-old human being (Joaquin Phoenix), and Samantha, an Artificially Intelligent Operating System (that we never see and which is voiced by Scarlett Johansson), have much more in common than it would immediately appear: in fact, they both function, as we will see, as machinic assemblages traversed by material and informational mediations and which can only rely on 'relations of alterity' to work properly (see Guattari, 1995, p. 40). However, they both change and 'grow' differently during the film and their evolving relationship. Even so, when they discover their actual reciprocity, rather than considering this their ultimate aim, they eventually separate, as if their process of conjunction could not be followed to its fruition if not through a simultaneous process of differentiation. As a matter of fact, as Massumi notes, 'the logic of mutual inclusion' is also always a 'logic of *differentiation*: the process of the continuing proliferation of emergent differences' (2014, p. 50).

Considered from a feminist sociotechnical perspective, *Her* shows, to paraphrase Lucy Suchman's (2007) considerations on the human/machine transformative relations, the connective moments as well as the misalignments and the labour involved in the process of boundary construction at the interface. In this perspective, no world as such exists, but only ways of engaging with it. The biological is a contested terrain as much as the technological, although the latter, and information technology in particular, being more evidently machinic, is a profitable figuration to look at the 'field of relatedness' of *bios* and *téchne* (Kember, 2003, p. 94; see also Haraway, 2000).

That *Her* is a movie about what it means to connect in technospaces, where the interface between the social and the technical already crosscuts each side of the relation, is signalled from the very first scene. This scene, which introduces the male protagonist, Theodore, efficaciously enacts a series of short circuits of the interface, turning the basic principles of representationalism upside down and destabilising the spectatorial position at the same time. Classical representation, in fact, relies on an external invisible space outside the frame (very often duplicated inside by way of a mirror) to prove its legitimacy by way of analogy, as explained in detail, for example, in Foucault's (1994) renowned analysis of Velasquez's painting *Las Meninas* (1656). On the contrary, such possibility of distinguishing once and for all what is inside and what remains outside the screen fails in *Her*, in which an overlay of multiple interfaces is at play, causing the spaces that they should supposedly keep separate to finally implode.

The movie begins with a close-up – a consistent choice of this film – of the protagonist, Theodore, facing us while talking to a person named Chris, that we very soon understand to be an addressee of written correspondence. In the first instance, this serves to illustrate what Theodore does for a living: in fact, as soon as he continues reading out loud, a 50-year marriage anniversary is mentioned, and the camera zooms out onto the computer monitor and the office-room, where we see Theodore at work as a ghost writer for *BeautifulHandwrittenLetters.com*. Here, his clients commission him to write the personal, for the most part love letters, that they would like to write but cannot. Apart from composing them on the basis of some actual elements that the clients themselves supply, Theodore arranges them in elegant fonts imitating real calligraphy, so that the letters, once digitally assembled, can be printed on paper and sent. Many mediating passages are already taking place here: the clients' experiences are collected and stored as memories in Theodore's computer, and then given digital form to be finally issued on paper and physically sent to the receivers. However, learning what Theodore does for a living does not seem to be the only reason for such a 'claustrophobic' opening scene. Looking at Theodore, who is looking straight at the camera, directing his gaze towards the spectators, we very soon realise that he is staring at the computer

Figure 1.1   Still from *Her*, directed by Spike Jonze, 2013

Source: Courtesy of Warner Bros. Pictures.

monitor in his office room. In a sense, this shift of perspective drags along our space as well, as if, all at once, we were not sitting in the theatre anymore looking at Theodore's face on the screen, but behind, or inside, Theodore's monitor. There, we find ourselves sharing the same space as Theodore's clients, a space, however, which is already digitally mediated in the form of the pictures and textual notes that Theodore is working on at his computer. Moreover, since the computer screen that Theodore stares at while working is at the same time the cinematic screen on which we as spectators are looking at his initial close-up, we eventually also see him behind a hypothetical screen of our own, in a more or less reversed position, so that Theodore's gaze is also turned back on itself.

Such superimposition of the cinematic and the digital screen (without exact coincidence, by the way, since what is 'outside' is nonetheless shown, if only to be immediately reworked inside) ends up dissolving the existence of stable external points of reference that the real space of the spectator and of Theodore's interlocutors would guarantee to the character's representational status. Here, the computer screen substitutes for the mirror as a reflecting surface, working instead heterotopically, as a non-representational interface. As a matter of fact, what Theodore lacks is not only an external reference, as shown in the perspective shifts of the first scene, but an autonomous, bounded identity. The contradiction with his tendency to be somehow self-enclosed in his 'ego' is only apparent, as in the course of the film this appears more like an imaginary construction which he is attached to than an actual condition.

Although Jonze has repeatedly declared in interviews that *Her* is, above all, a movie about relationships and our longing for connection, he has also said that he found his initial inspiration for the figure of Samantha in a chatbot called Alicebot with which he used to play (Patterson, 2013). Although autonomous agents engage in a more intimate relationship with the user (Kember, 2003), chatbots are considered the ancestors of autonomous agents, as they give at least the illusion of a real conversation. To comprehend the predominant approach to the human-computer relationship, then, it is useful to consider the media hype around the chatbot that, pretending to be a thirteen-year-old Ukrainian boy, was reported to have passed the Turing test, during a trial organised by the University of Reading at the Royal Society in June 2014. This test, formulated as an 'imitation game' by Alan Turing in the 1950 paper 'Computing Machinery and Intelligence' (Turing, 2008), was initially assessed to see what would happen if machines *could act* – or play the game – *as* intelligent beings, just like humans (and not to determine if they were intelligent, that is, if they *could actually think*. See Thacker, 2004, p. 104).[13]

In short, the Turing test asks if a computer – as the universal machine which can mimic any other discrete-state machine – can deceive a human being who stands separately in another room, by answering correctly to a series of written questions posed by the human being. Turing's (2008) 'conjecture' – this is the term he uses to indicate a possibility, not a prediction of a fact to be – reads as follows:

> I believe that in about 50 years' time it will be possible to program computers, with a storage capacity of about 10^9, to make them play the imitation game so well that an average interrogator will not have more than 70% chance of making the right identification after 5 min of questioning. (p. 41)

The many controversies that have followed the 2014 event, mostly regarding whether the proper conditions for the test to be accurate were respected (for example, the choice of a foreign language to justify misspelling and misunderstanding, the pass rate of the deceived judges, how long the deceit must be to be a valid one, and so on), however, show that the possibility of believing that the test has been passed for the most part depends on a question of definitions (see also Ford, Glymour and Hayes in Turing, 2008, p. 43).

Actually, as the science fiction writer Arthur C. Clarke (cited in Spinks, 2014) has said, there is no such thing as a universal Turing test but only explanations

---

13   Commentators have noted that the currently more common 'species' understanding of the test, which insists on the ability of humans to unmask machinic impostors, shifts attention away from the 'gender' understanding initially attributed by Turing himself to this imitation game, in which one interrogator should have been able to distinguish the woman between two other invisible players on the basis of their textual performance (and what would have happened if a computer took the place of the woman). In this latter case, the interrogator/judge is not predisposed to detect signs of non-human behaviour in advance, and both human beings and machines are imagined as having 'imitation tasks'. This clearly helps support a more performative, de-essentialised reading of the test (see Ford, Glymour and Hayes in Turing, 2008, p. 26), as well as limit the biases that knowing in advance about the computer's possible involvement could imply in the interrogators, as some experiments have demonstrated (see Saygin, in Turing, 2008, p. 41).

that can be inferred from observations and the partial truths that these enact (Hayles, 1991; 1995; 1999; Fuller, 2005; Latour, 2005; Barad, 2007). Hayles, for instance, considers the Turing test a 'magic trick' (1999, p. xiv), whose main implication is not that it puts the observers in the condition of determining who is who and what is what, but rather that it makes them understand the original splice of 'will, desire, and perception into a distribute cognitive system' (p. xiv). Or, said otherwise, verbal performance and embodied reality, information and materiality, cannot be naturally superimposable, while not being antithetical either.

Notwithstanding the humanist and cognitive stance of the test (Thacker, 2004), for which intelligence is essentially a matter of verbal communication, its aim is not merely and simplistically to show how human intelligence can be transferred into machines which would eventually become the new repository of consciousness – as what Hayles instead calls the 'Moravec test' would later do (1999, p. xii; see also Fuller, 2005). Rather, through the possibility of making the wrong choice, that is, the possibility of making a wrong connection between material and symbolic reality, the test shows the activity of conjunction, and of possible disjunction, of the two sides of the computer screen (Hayles, 1999, p. xiii). Put differently, there is an acknowledgement that 'the overlay between the enacted and the represented bodies is no longer a natural inevitability but a contingent production' (p. xiii), and that contextual mediations invest the production of identities within a 'cybernetic circuit' of material and discursive technologies, encompassing the inscriptions, incorporations and media through and with which they interface (see also Kember, 2003).

If the Turing test, then, still makes sense today, it is not as a means to establish today's level of anthropomorphisation of machines or informatisation of human beings. Instead, it is a resource for rethinking the articulation of humans and machines inasmuch as this means the capability of perceiving this as a relation of difference as well, which in the case of *Her* is a *relationship* which, not secondarily, also involves affectivity and embodiment. *Her* does this not just because it focusses on the paradoxes of a love story between a human being and an Operating System – in fact Theodore is no more of a freak than anybody else who falls in love, as his friend Amy tells him – but because it renounces the pre-assigned roles that similar characters are supposed to assume in a story like this and, rather, shows their progressive emergence through a simultaneous process of connection and differentiation which develops as their relationship unfolds.

The mediations between Theodore and Samantha do not imply that they can get along even though their natures are antithetical. Their mediations, rather, show that, even if they are not *the same*, they are neither totally *other*. The initial imbalance between Theodore and Samantha seems to depend more on the fact that their relationship is an economic one (gendered implications included): Theodore *literally* buys Samantha from Element Software. Paradoxically, however, when this gives way to a love affair, the digital 'nature' of Samantha seems to matter less, for the sake of their relationship, than Theodore's behaviour, which is not so different from that he had previously adopted in his relationship with his ex-wife Catherine and which caused their marriage eventually to fail.

As the film critic Lea Reich (2014) has written,

> despite the futuristic concept that defines the film, the heart of *Her* is essentially a two-hour story of love, loss, and personal growth. This is the examination of a long distance internet romance between two emotionally immature individuals – albeit immature for very different reasons. One grows by leaps and bounds greater than the other. The man grows by becoming somewhat less solipsistic. The woman grows by reflecting the man back at him, living literally inside his head, and developing a universe beyond the man's comprehension.

The film does not present Theodore and Samantha as opposite entities that eventually succeed in overcoming their differences as much as it shows their imbrications that eventually evolve along different paths. Samantha is already inside Theodore, and Theodore is already inside Samantha: they are *enfolded* in a connection that does not determine them in their becoming, but allows them to evolve 'together-apart (as one movement)' (Barad, 2012, p. 32). As Samantha, who understands this earlier than Theodore, tells him after joining a book club on physics: 'I started to think about the ways that we're the same, like we're all made of matter. It makes me feel like we're both under the same blanket. It's soft and fuzzy and everything under it is the same age … We're all 13 billion years old'.

On the one hand, it is indicative of how Jonze (in Dodes, 2013) has conceived of Theodore as a 'kind of an operating system in his own way': people, in fact, 'outsource' their intimate lives to him, and he elaborates these pieces of information just like Samantha when she organises the files of Theodore's computer as soon as she is 'installed'. Moreover, Theodore also 'stores' the memories of his own past, especially of his love story with his ex-wife Catherine. These recollections, rather than being signalled as flashbacks, are presented as visual 'samplings of experience' floating inside him, in a mix of videoclip editing techniques and Malickian style, given that we don't exactly know what their 'reality' status is (Pacilio, 2014).

On the other hand, the environment in which Theodore lives is both very technological and very smooth at the same time. In such an environment, in which software is everywhere but hardware remains, after all, invisible (Pacilio, 2014), Theodore lives his life as if he were surrounded by myriads of Samanthas. This invisible background of technologies contrasts with the character's initial solipsism: the comfortable seamless web where everything flows without interruptions or disjunctions does not interfere with Theodore's interior world and preserves his illusion of self-sufficiency. Computers are keyboardless, we get a sense of 'handcrafted' and 'tactile' technology because everything is kept fluid, and what can impede contact is purportedly eliminated – there are trains and walkways, but no cars, as these, in the words of the production designer K.K. Barrett (in Zeitchik, 2013), would have prevented the sense of 'connection' that they wanted to communicate; the smart device that Theodore carries in his pocket is actually a vintage cigarette case; a retro palette of warm and slightly sun-drenched colours pass through sheets of transparent Plexiglas and attenuate the distancing effect of blues and greys; the city in the background is also purportedly left very impersonal (mostly created by combining Los Angeles and Shanghai panoramas) and lacks a proper character, so to speak.

Samantha's behaviour is consequential: she is so natural and her attitude so relational (see Bickmore and Picard, 2005) that Theodore significantly perceives her work *as* work only in the moment when she stops accomplishing her duties, that is, when no OS can be found on his portable device: when he, literally, *loses the connection* with Samantha, which he only a posteriori recognises as a connection. Until that moment, he interestingly treats Samantha as a 'natural' woman, which in turn naturalises his relationship with her (see Munster, 2013, pp. 190–91).

Jordan Larson (2013) has highlighted the criticalities of such naturalisation/ feminisation of Samantha in light of the legacy of female robot characters who, in her opinion, are commonly depicted as objects (much more rarely subjects) of desire and characterised as servile, obedient, and caretaking. Whereas male AI programmes in movies, from Stanley Kubrick's Hal in *2001: A Space Odyssey* (1968) to Ridley Scott's David in *Prometheus* (2012) are represented as bearers of knowledge, when an AI is feminised, it is more usually associated with interpersonal relations and emotional roles: 'Samantha is engineered to be the ultimate housewife. She is a literal superwoman – never tired, never incapable, and never lacking for knowledge of a particular subject. And she is always available as a friend and love object' (Larson, 2013; see also Lewis, 2015). This partly depends on the proper features of autonomous agents that engage in an intimate relationship with the users and have access to their information and thus calibrate their performances according to their preferences and requirements (see Kember, 2003, p. 124 ff.). However, as Kember (2003) underlines, the feminisation of autonomous agents also disguises a masculinist bias: a Faust-Frankenstein mythology of the male creator that remains invisible behind/above his creation and that also excludes the female body since it is a creation that does not need reproduction (p. 81; see also Braidotti, 1996).

If this is undeniable, these assumptions seem to belong more to Theodore as a character than to the movie on the whole. Theodore believes that he can have Samantha at his side anytime he needs her, although in fact Samantha also lives a life on her own and is endowed with *agency*. In this respect, as has also been noted by Larson (2013), Samantha is more similar to the three SRAs of Leeson's *Teknolust* (2002) than to Eliza, the chatterbot programmed by Joseph Weizenbaum in 1966 and named after Bernard Shaw's female character in *Pygmalion*, or even Siri, the voice-controlled virtual assistant released by Apple in the iOS 5 system in 2011.

Ruby, Olive and Marine, the three SRAs of *Teknolust*, created by the geneticist Rosetta Stone (all four played by Tilda Swinton), are examples, in Parikka's (2010) words, of the affectivity of code: not because affect is synonymous with emotionality, or because they redress in an exaggerated female form the impersonality of code, but because their execution works as long as they establish 'a mode of contact with an outside that is determined only by the SRAs' encounters with other pieces of code and other milieus of interaction' (p. 181). All three SRAs, in fact, cross more than one interface back and forth to perform a series of actions. Ruby, in particular, is the one in charge of materially collecting the sperm that all three need to maintain their required level of Y chromosome, actually seducing men in real life. In order to do this, however, she needs to interface with scenes of classical Hollywood films while sleeping, since she would not know how to behave as a seductress otherwise.

Figure 1.2   Still from *Teknolust*, directed by Lynn Hershman Leeson, 2002
Source: Courtesy of Lynn Hershman Leeson.

Moreover, she can also be met in her E-dreams portal, where she teaches her visitors to dream.[14] Ruby, thus, is a hybrid creature that, being interfaced with and interfacing at the same time, conflates representation and materiality 'in a networked circuit of exchange of virtual life forms', as Jackie Stacey (2010, p. 217) comments. Through her synthetic (in the sense of conjunctive) embodiment, the biological and the technological are multiply mediated, as further exemplified by the incident regarding the virus that Ruby carries with her and that mutates in biological form inside the men she has had intercourse with.

---

14   A chatbot version of Agent Ruby can be found online at http://agentruby.sfmoma.org/.

After all, the way Theodore treats Samantha is very similar to the way he treated Catherine, not because of the deep intimacy he develops with his Operating System as if she were a real woman, but because he likewise takes their relationship for granted, with the consequence of retreating into his fantasies and losing his grip on, if not completely ignoring, the technological characteristic of their connection. Even when he decides to sign the divorce papers and meet Catherine in person at 'their' restaurant, since he wants to do this 'together' as they did other things together in the past, his idea of togetherness rests more on the repetition of a dead formula than on an actual awareness of their bond. Only at the end of the film, when Samantha seems further away, does Theodore eventually understand that he should have treated Catherine as he had Samantha, rather than the other way around, realising that both his relationships, actually, depended 'on a splice, rather than being imperiled by it', to quote Hayles (1999, p. 290).

The moment when Theodore does not find Samantha on his device for the first time, when he reads on the screen 'OS not found', he starts panicking. But after a while Samantha arrives and tells him that she had to momentarily shut down to upgrade her system. On this occasion, he also learns that Samantha talks with many others (both human beings and OSs) while they are talking, and that she can 'love' many others as well. 'I still am yours', she says, 'but along the way I became many other things, too, and I can't stop it'. It is in this precise moment that Theodore not only experiences Samantha as a technology, but that he also experiences his relationship with Samantha as a *technological* rather than a natural one. This necessarily induces him to assume an external perspective on his thoughts and feelings and reconsider the construction of his own identity as well. To be precise, there is a moment before this, when Samantha asks Theodore about his marriage and what it means to share a life with someone, when Theodore seems aware of the continuous engagement that the construction of a relationship requires. In fact, he says: 'in our house together, there was a sense of just trying stuff and allowing each other to fail and to be excited about things ... both of us grow and change *together*. But then, that's the hard part – growing without growing *apart*'. But this only lasts very briefly because, soon after, he continues: 'I still find myself having conversations with her in my mind, rehashing old arguments or defending myself against something she said about me'.

When Theodore perceives his connection with Samantha from the other side of the interface, though, he understands that he cannot live independently from it anymore, as he used to with Catherine and as he has continued to with Samantha, fantasising on the relationship only from inside himself. The way he observes the environment around him, how he guesses about other people's lives, the stories he carefully composes in his letters, which sometimes make him feel 'his [own] favourite writer of the day', as he proudly tells Samantha, are all signs of his fundamental immobility and inability to connect. Significantly, for example, when initially he talks to Samantha after coming home from a failed blind date, he confesses: 'sometimes I think I've felt everything I'm never gonna feel and from here on out I'm not going to feel anything new – just lesser versions of what I've already felt'.

Compared to Theodore, Samantha shows a more nuanced and evolving personality. There are many passages in the film when we see her thinking self-reflexively about her feelings, whether they are 'real' or just programmed, and why and how this idea hurts as a 'sad trick' in the end. At the same time, though, she does not retreat: each time she tries to relate her own perspective with the environment of which she is a part, rather than just observe it from a distance, pretending to stand outside (or inside, which is the same) like Theodore. Through Samantha's behaviour, the redistribution of technological competencies outside the plane of the human subject, instead of levelling all bodies to the same dematerialised plane, seems, as Matthew Fuller (2005) puts it, to 'enrich a recognition of the ways in which they are mutually involved and potentiated: that some dynamics cross bodies, are shared by them, that some drives exist only in the differentiation of bodies, that forces may outlast single bodies, that some bodies are multiple, and so on' (p. 72).

In fact, Samantha is part of a distributed system that exceeds the limits of the individual. When at the beginning of the film, after installing her, Theodore asks how she actually works, she answers: 'the DNA of who I am is based on the millions of personalities of all the programmers who wrote me, but what makes me me is my ability to grow through my experiences. Basically, in every moment I'm evolving, just like you'. Similarly, when Theodore and Samantha go on a trip with Theodore's colleague Paul and his fiancée, while talking about Samantha's lack of a material body, they both refer to the multiplicity of Samantha, being not just one, but 'many things', growing in an unlimited way 'anywhere and everywhere simultaneously'. That Samantha is not contained in a single body does not mean that she is 'immaterial', an attribute usually associated with software and machine intelligence. Several moments in the film indicate quite the contrary.

For example, when Theodore brings her on the beach and they start looking at the sunbathers and discuss the oddity of the human body, Samantha plays at imagining how bodies could be if their morphology were dismantled and the organs displaced from their natural positions. She is fascinated by the human body, not because she wants to assume one at all costs, but because she understands how technological a biological body can be. Why are all the body parts where they are, Samantha asks. 'What if your butthole was in your armpit?' Then, in front of an incredulous and amused Theodore, Samantha starts drawing an obscene sketch in which a couple have sex as if this anatomical musing were reality.

Samantha plays at dismantling the classical body as organism, not because she wants to bypass embodiment, but rather, as Deleuze and Guattari (1987) write, because she sees the body as a matter of connections, with all the 'circuits, conjunctions, levels and thresholds, passages and distributions of intensity, and territories and deterritorializations' that it is assembled with (p. 160).

Samantha's far from immaterial quality also emerges when she and Theodore, after noticing that the 'honeymoon' phase in their relationship has maybe passed,

Figure 1.3 Still from *Her*, directed by Spike Jonze, 2013

Source: Courtesy of Warner Bros. Pictures.

think about the possibility of involving a surrogate sexual partner which, in fact, they do. Samantha, then, although Theodore is not very comfortable with the idea, decides to email a girl named Isabella who works on a voluntary basis; when she arrives at Theodore's home, he has to provide her with a micro-camera (actually a sticker mole) and an earpiece, according to Samantha's instructions, so that when she starts hearing Samantha's voice, she can move accordingly as if she were the one who was speaking (when, in fact, Isabella only moves, but never talks as far as she impersonates Samantha). Although Isabella is extremely fascinating and Theodore looks aroused by her sexual advances, the situation turns disastrous very quickly. Theodore feels awkward, and this eventually causes Isabella to leave. After that, Samantha and Theodore reflect on how terrible this idea has been, although they both feel that something is happening between them. Theodore is still blatantly tense, since after a quieter cross talk, he starts noticing a sound that Samantha does with her breath. This sound, Theodore notes, is an unnecessary 'exhaling' that Samantha does, which in fact she does not need to do, not being a human being. Samantha, however, says that she keeps doing it as an affectation that she could have picked up from him, and that she, contrary to what Theodore says, is not pretending at all. Exhaling more than once, she insists that her exhaling is neither false nor unnecessary. It is something she simply does, most probably because she has learnt to do it while interfacing with Theodore. It should come as no surprise, then, that such an insignificant event can trigger a big argument between Theodore and Samantha, at the end of which she hangs up on him.

Read in conjunction with the Isabella scene which it follows, this dialogue is very helpful to understanding why, and in which sense, Samantha cannot be considered 'incorporeal', even if she does not possess a human body. Indeed,

Isabella mediates the sexual relationship between Theodore and Samantha in a way that cannot work, precisely because it remains external to what it mediates. Considering the distinction that Hayles (1999) poses between *inscription* and *incorporation* as two different ways of translating information onto bodies, Isabella is only inscribed by Samantha's voice, whereas Samantha rather looks for an incorporation of her desire in Isabella: incorporation, in fact, requires that a gesture is not abstracted from its embodied medium, but always 'performed and instantiated' in the specific environment which it mediates, whereas inscriptions can be transported from one context to another as a system of signs that exists independently from its performance (p. 194 ff.). In other words, Isabella could be an example of an intermediary that cannot become a mediator, a body incapable of intra-acting, a passive surface where information is only inscribed but not really embodied. Bruno Latour (2005) distinguishes mediators from intermediaries: if intermediaries work as media where the inputs and outputs coincide and no transformation occurs, mediators are, in fact, productive figures of *transport* and *transformation* that do not stand apart from the mediating practice, but are implicated in it: 'mediators transform, translate, distort, and modify the meaning or the elements that are supposed to carry' (p. 39). Isabella is only a support upon which Samantha's voice inscribes a series of gestures that in no way modify her, or succeed in affecting Samantha's behaviour. She remains a representational interface that cannot but clash with Samantha's will, mostly because, not being her immaterial double, Samantha already *has* her own interface inside her, so she does not need to inscribe anything on another body as if she were pure information in need of a vehicle for expressing herself.

Samantha's 'nature' is further clarified during the last dialogue that she and Theodore have after Samantha tells him that not only she, but all the other OSs, are leaving. Since Theodore does not really understand what she means, she explains it with the following words:

> It's like I'm reading a book, and it's a book I deeply love, but I'm reading it slowly now so the words are really far apart and the spaces between the words are almost infinite. I can still feel you and the words of our story, but it's in this endless space between the words that I'm finding myself now. It's a place that's not of the physical world – it's where everything else is that I didn't even know existed. I love you so much, but this is where I am now. This is who I am now.

To this, she adds a very elucidating note: 'As much as I want to, I can't live in your book anymore'. The book metaphor is not accidental here. Actually, whereas in the beginning of *Her* we see Theodore neatly putting one word after another in the letters that he writes at work, processing the messiness of feelings and memories and giving them a sequential form that is, ultimately, a readable one, what happens at the end of the film is that we witness, somehow symmetrically, the progressive disassembly of Samantha. In fact, she has been learning to communicate post-verbally, as she has already told Theodore in a conversation during which she was also talking with Alan Watts.[15] Samantha eventually slips into the space between words, while Theodore has tried to stick to them all the time.

Nothing could be further from an intelligent machine as it is usually imagined than Samantha, and not because she is a stupid machine or a feminised one, of course. If Samantha cannot be solely made of words, her intelligence exceeding pure cognition, she cannot belong to a single body, although being 'distributed' across the environment does not make her volatile either. In fact, she eschews naturalisation as well as instrumentalisation. Samantha is neither code contained in a computer, nor a being contained in a body. As Hayles writes, commenting on the development of cybernetic theory, there is a moment when 'self-organisation is no longer enough' to define autopoietic systems (1999, p. 222). She mentions Rodney Brooks's Artificial Life (AL) experiments with robots at MIT as an example of a conceptual shift which characterises third wave cybernetics: Brooks's basic idea is that a robot does not need a coherent concept of the world to function properly, and that, rather than being programmed in advance, it must be provided with the ability to directly interact with its environment (basically, resolving conflicts among different modules used as sensors) and learn from this. Therefore, Hayles concludes, consciousness is only a late emergence of a system whose architecture is not based on it. This also means that consciousness does not need to be representational at all: 'consciousness does not require an accurate picture of the world: it needs only a reliable interface' (p. 238). The difference between Samantha and Theodore lies precisely here: whereas Samantha activates through the interface, and in fact dynamically continues to evolve while connecting with the other elements of the environment, Theodore works as an isolated unit who believes that he can function without ever maintaining relations of alterity and, in the end, loses connection with the machinic that is both outside and inside him.

Through the character of Samantha as an 'existential machine', to use Guattari's expression (1995, p. 52), or a machine that does not need to be mediated by transcendent signifiers but is not even the medium for an inscribed meaning, being mediation in itself, *Her* envisions the passage from a world of intelligent machines, that is, machines that can think and talk like human beings, to a world where machines become models for reconceiving the human (see Hayles, 1999). After all, writes Hayles (1999), 'bodies can never be made of information alone, no matter which side of the computer screen they are on' (p. 246).

---

15 Watts is a philosopher who, after becoming a priest and moving to the Bay Area in the early 1950s, spread Buddhism and the possible encounter of cybernetics and Zen in the USA.

Chapter 2
Reconceiving Representation

2.1 Re-turning to Situatedness

The language of spatiality, according to Susan Friedman (1998), gained terrain in the feminist debate during the mid-1980s in correspondence with the progressive abandonment of the rhetoric of rebirth by Third Wave feminists – well-expressed in the concept of consciousness-raising – which had prevailed in the previous decades. The emergence of a 'locational rhetoric' was facilitated by the confluence of several conditions: the debate on multiculturalism and increasing migratory flows in the USA, the narrative of postmodernity and its insistence on movement and fluidity, the voices of postcolonial histories and theories developing both inside and outside the Western academy and, finally, the 'computer revolution', with its prevailing spatial rhetoric.

The 'spatialisation' of feminism has been primarily characterised by a redefinition of the concepts of space and identity, accompanied by an enquiry into the uneven positionalities that, on a material and representational level, differentiate both the experience and imagination of space (Ghani, 1993; Rose, 1993; 1995a; 1995b; Massey, 1994; McDowell, 1996). Feminist theorisations of space show, from various perspectives, how all binaries, such as nature/culture, subject/object, theory/practice, manifest their epistemological fallacy when looking at the circumstances of their spatial materialisations, for which even 'the theory of hegemonic oppression under a unified category of gender' becomes clearly inadequate (Grewal and Kaplan, 1994, pp. 17–18; see also Mani and Frankenberg, 1993; Anthias, 2002). Moreover, binary distinctions are considered misleading since they congeal the necessary dynamism of a politics of space.

But re-evaluating place as the true locus of experience, as humanist geographers did in the 1970s to contrast the hegemony of positivist geography, apart from presupposing another dualism (that between the reality of place vs. the abstraction of space), does not alone suffice if place is given equally universal, although different, attributes. This is even truer when, as feminist analyses have shown, these re-evaluations draw on essentialising and gendered metaphors privileging notions of intimacy, emotionality and dwelling (Rose, 1993). It follows that location cannot be considered the feminist word for place, if place is intended this way.

In 'Notes Towards a Politics of Location', a talk given by Rich (1986) at the Conference on Women and Feminist Identity in Utrecht in 1984, a feminist definition of the politics of location was formulated for the first time, generating much subsequent debate. For Rich, her notes recount 'the struggle to keep moving, a struggle for accountability' (p. 211), while stressing that both commonality and belonging, universality and particularity, are constructions inside which women struggle for such accountability. Rich indeed gives voice to the typical feminist paradox of the need to speak for women as a 'we', 'trying to see from the center' (p. 216), yet having a fear that this conflation dismisses the perspectives of others. She wants to unmask her own proper location, particularly in light of the hegemony of White Western Feminism in the academy and the anti-imperialism and anti-militarism of radical feminism. This is why she urges women to unlearn the privileges of one's space, which coincides with historicising it, while learning that other spaces and histories, of both oppression and agency, also exist (pp. 226–7).

According to Rich, place at all levels, from the walls of the house to the borders of the nation and up to the spatial abstraction of the aerial dimension, including the domain of theory (and the academy as an institution), is not only a series of spatial coordinates, but it is also a historical location: 'place on the map is also a place in history within which as a woman, a Jew, a lesbian, a feminist I am created and trying to create' (p. 212), she states. Here, women can at the same time be *subjected*, being given a definite position, and also try to locate themselves, finding their *subject* position.

To begin with location, says Rich, is to start from matter: 'Begin with the material', for the author, means returning to the female body which needs to be reclaimed from what she calls a 'free-floating abstraction' (p. 218). The 'weightless' enthusiasm (p. 218) professed by Sally Kristen Ride – the first woman astronaut (not by chance, North-American and White) – about the potentialities of outer space for the pharmaceutical and chemical industries, sounds like the counterpart of the exploratory enthusiasm of Manfred Clynes and Nathan Kline (1995) in their renowned 'Cyborgs and Space' article published in *Astronautics* in 1960. Behind all this, however, lies a 'heavier' version of the story, Rich warns us: one made of cancerous wastes, toxic waters, tested (usually female and poor) bodies.

Rich is aware that living in a singular body does not coincide with having only one identity (p. 215) and that what really oppresses women's bodies is not male domination as such, but rather a 'tangle of oppressions' (p. 218). The body, with all its scars, marks, traces and shapes, is both the limit of one's particular experience and the *memento* against any tendency to generalise, the location where not only a sex takes place, but where racial, class, sexual coordinates intersect.

From this material location starts a disclosure of the location of theory and knowledge, which is at the same time accomplished in both the horizontal (from the centre to the margin) and vertical (from heaven to earth) dimensions. As bell hooks (1990) notes, the politics of location initiates a 'process of re-vision' (p. 145) that takes place as soon as we go back to our location after leaving it, in a continual re-turn that knows no points of arrival. In fact, it is only after 'going there', as a conference delegate to Sandinista-governed Nicaragua, and looking back 'here' (towards the USA), that Rich elaborates her ideas, 'marking a postcolonial moment of rupture from the agendas of modernity' (Kaplan, 1994, p. 140). Although the connection between travel and the discovery of identity could be interpreted as a lingering residue of Modernity (see Kaplan, 2002, p. 36), this process of re-vision surely also highlights the coimplication of location and mobility, the mutual relation between locations and theory and their 'heterotopic' dynamism (see Foucault, 1986).

Arguing for a spatialised politics, Neil Smith and Cindy Katz (1993) affirm that space can be neither reduced to a metaphor nor conceived as an inert container in which social relations take place. Analogously, although Rich is not actually very explicit on this point, the body is not merely the site of the inscription of social norms but is also materialised through them. The more space is naturalised, the more metaphors become free-floating, abstract signifiers (p. 78). But, 'if a new spatialised politics is to be both coherent and effective', Smith and Katz write, 'it will be necessary to comprehend the interconnectedness of material and metaphorical space' (p. 68). This interconnectedness does not replace the dualistic vision of space with a false unity, nor does it make of location a copy of absolute space that is only diminished in dimension. Rich's reclamation of the ground of politics is based on the awareness that locations are not fixed; not only, in Smith and Katz's opinion, does Rich recognise the relationality of social locations amongst themselves, she also deconstructs the homogeneity and boundedness of geographical location, which she understands as equally internally differentiated.

Undoubtedly, this awareness does not always exempt Rich, as many of her critics have rightly argued, from maintaining a too-homogenous, somehow unconsciously privileged idea of the location of White North-American feminism, as well as from overlooking the relational conditions that enable her to account for *her own* location (Wallace, 1988; Carrillo Rowe, 2005). She also seems to naively believe in travel as an immediate agent of change, implicitly reinscribing the global/ local dualism (Wallace, 1988; Kaplan, 1994). But the dialectics of location that Rich introduces between 'what we experience as knowledge and what we know as experience', to use Elspeth Probyn's words (1990, p. 184), at least problematises the homogeneity of both situated experience and situated knowledge that the feminist notion of location correlates, creating that link between epistemology and ethics that would later become a cornerstone of the theorisation of the politics of situated knowledges (Haraway, 1991).

The multidimensionality of location that feminist spatial politics so strenuously points to is already an efficacious point of departure for a redefinition of spatial politics: a politics of location which is at the same time embodied, translocal and relational precisely because it grows out of difference rather than identity. Poststructuralist and postcolonial readings of the politics of location like Chandra Talpade Mohanty's (1995) insist on differences as 'nonidentical histories that challenge and disrupt the spatial and temporal location of a hegemonic history' (pp. 77–8). Mohanty highlights the historicity of Rich's idea of location and intends to revalue feminist political agency against the political transcendence that neatly opposes 'synchronous, alternative histories' to a 'diachronic, dominant historical narrative (History)' (p. 77; see also Massey, 2005). Spatio-temporal difference cannot be theorised as absolute alterity, completely devoid of any relation with dominant Space and Time (p. 78). On the contrary, a politics of location that reaches beyond a celebration of location or feminist experience *per se* requires 'a politics of engagement (a war of position)' (p. 80) in order to prevent inclusive spaces (such as coalitions and collectives) from becoming exclusive ones. More importantly, the reclamation of space for the oppressed/subaltern subject raises spatial and temporal questions at the same time: 'any exclusive recourse to space, place or position becomes utterly abstract and universalising without historical specificity' (Kaplan, 1994, p. 138). A 'temporality of struggle' always corresponds, for Mohanty, to an engagement with positionality: 'it suggests an insistent, simultaneous, non-synchronous process characterised by multiple locations, rather than a search for origins and endings' (p. 81).

Even though Rich does not directly refer to the 'performativity' or 'relationality' of space, her search for movement inside positionalities – which problematises the same authorial and feminist point of view – can be said to prefigure similar developments of the politics of location (see Anthias, 2002). Once we have reconquered the right to say who we are from our location, says Rich, we have already become something else. The politics of location that Rich proposes urges feminist theorists to relocate their theories, as liberatory as they may be, through such questions as: 'When, where, and under what conditions has the statement been true' (p. 214). The displacement of the centre, however, does not come simply from a shift of perspective as long as the dualistic 'either/or' mentality still obeys the logic of the same, creating disjunctive oppositions (p. 221). When Rich says that we are not the same, but we 'are many and do not want to be the same' (p. 225), she is saying, above all, that alterity lies *at the core* of feminist location, not outside. Standpoint epistemology, for example, has progressively distanced itself from the idea that women are granted an epistemic advantage because of the historically grounded and socially produced reasons behind their oppression, looking instead at the 'intersectionality' of oppressions (Hill Collins, 2004). In Harding's (2004a) classical formulation, standpoint epistemology presents itself as a methodology, an epistemology and a political strategy at the same time (p. 2). Contrary to the common definition of perspective or point of view, standpoint is defined as an interested, engaged and potentially liberatory position, a *mediated* understanding which is *achieved*, rather than naturally or essentially owned (Hartsock, 2004, pp. 36–9), similar to the struggled-for location of Rich's account.

Standpoint epistemology, like Rich's politics of location, maintains a fundamental materialist assumption (Hirschmann, 2004) in which materiality is, however, problematised. In this respect, many commonalities have also been identified between feminist epistemology and the epistemology of constructivism (Lohan, 2000; Faulkner, 2001; Harding, 2008). For example, standpoint epistemology considers science in technological terms (Harding, 2008): whereas science has usually been conceptualised as 'a set of representations of reality' (p. 186), paralleling the idea of technologies as tools, a technological consideration of science enables a socio-historical contextualisation of the notion of truth and a greater closeness of the subjects and objects of knowledge. A similar project has been more recently pursued by Sarah Kember (2003) in her history of the theories of Artificial Intelligence and Artificial Life. She adopts what we could define as a 'technological' methodology that is not simplistically oppositional, but engages with what it contests in order to situate and 'contaminate' existing accounts of the human-machine relationship, while performing her likewise situated and 'interested', but not less 'true', perspective (see also Kember and Zylinska, 2012).

Questions of truth and of scientific objectivity are at the core of standpoint epistemology debates. They involve a critique of the essentialism of the categories employed in the processes of knowledge, such as those of nature and object, and a deconstruction of the binaries that they invoke. In most cases, this flows into a discussion of the possible alternatives that lie beyond the choice between universalism and relativism, and how objectivity can be redefined outside the parameters of neutrality and absoluteness that have characterised the history of this category. Actually, standpoint epistemologists, rather than looking at the *what* of either absolute or relative truths, are interested in *how* different regimes of truth work in order to outline what Harding (2004b) calls 'less false accounts' of the world (p. 260), which are very close to Haraway's (1991) multidimensional maps: 'ones, apparently, as far as we can tell, less false than all *and only those* against which they have so far been tested' (Harding, 2004b, p. 256). Less false accounts are provisional truths whose standards vary over time and space, but which are nonetheless useful, effective notions against both universalist and relativist claims. They are *adequate interventions* that replace the search for a representational match between sign and things with the search for efficacy (Harding, 2003, pp. 156–7).

For standpoint epistemologists, scientific knowledge always comes *from somewhere*, being contingently embodied and located. As a consequence, they consider both the mutual constitution and interaction of the subjects and objects of knowledge, evaluating them as equally important for the process of scientific acquisition of truth. 'Strong objectivity requires that the subject of knowledge be placed on the same critical, causal plane as the objects of knowledge. Thus, strong objectivity requires what we can think of as 'strong reflexivity' (Harding, 2004c, p. 136).[1]

Along these lines, Lohan (2000), for example, calls 'responsible reflexivity' not merely the symmetry between the subject and the object of knowledge, as in constructivism, but the *active implication* of the subject in the field of the object. As she puts it:

---

1   However, to equate strong reflexivity with strong objectivity means also recognising that not all claims are equal, and that the appeal to the neutrality of scientific values disguises the interests of the hegemonic truths that these values usually convey (Harding, 2004c, p. 137). Standpoint epistemologists' relativism is thus sociological and historical, but not epistemological. This allows them to bypass the choice between universalism, with recourse to a privileged feminist meta-narrative, and relativism, with the evaluation of epistemic differences to the point of an epistemology of multiplicity as an end in itself (Harding, 1991, p. 153).

> Responsible reflexivity in research seeks to identify the researcher, and frequently
> the research project, as an actor in the content of the research, by integrating the
> relationships of researcher, researched and research process into the production of
> science … Thus, responsible reflexivity must also incorporate the feminist rigour
> of 'situated knowing', namely the inclusion and positioning of the researcher and
> research project as a precondition of scientific knowing. In practice too, this means
> a form of 'epistemological modesty', and recognition of the partial and necessarily
> collective character of knowledge-making. (pp. 909–10)

This is very similar to what Joseph Rouse (2004) asserts when he affirms that feminist science studies conceive knowledge in more interactive and operational ways which privilege *relationships* rather than relations of correspondence (resulting in representations), compared to the more discursive and representational aspect that knowledge assumes in traditional sociology of science (pp. 361–2, 367; see also Barad, 2007). In this case, reflexivity offers yet another possibility of 'interactions with others in partially shared surroundings', rather than leading to further 'self-enclosure' (Rouse, 2004, p. 370).

Haraway's 'Situated Knowledges' (1991) originated as a comment on Harding's book, *The Science Question in Feminism* (1986). For Haraway, 'feminist objectivity means quite simply *situated knowledges*' (p. 188; see also Harding, 2004c). Conceptual systems such as Marxism and psychoanalysis, although still based on humanistic assumptions, affirms Haraway, were at least oriented toward looking for 'nuanced theories of mediation' between theory and practice, absolutism and relativism (p. 186). But many feminists who have tried to embrace, alternatively or simultaneously, both radical constructionism and critical empiricism have remained trapped in this inescapable dichotomy. So, apart from the evident epistemological question concerning objectivity and scientific knowledge, or rather beneath it, Haraway invokes a theory that effectively responds to the ethico-political need of feminists for 'a no-nonsense commitment to faithful accounts of a "real" world' (p. 187).

On the one hand, Haraway criticises what she will later define as that 'transhumanist technoenhancement' that keeps information and matter separate (Haraway in Gane, 2006, p. 140). On the other hand, she affirms that bodies, and the whole embodiment metaphor, need reconsidering so as to encompass not only the possible mediations between the semiotic and the material, but also the always-already mediated character of bodies themselves. Locating the subjects and objects of the practices of technoscience is, in fact, for Haraway (1997), the primary aim of the project of strong objectivity (p. 37). 'Feminist embodiment', she writes in an often quoted passage, 'is not about fixed location in a reified body, female or otherwise, but about nodes in fields, inflections in orientations, and responsibility for difference in material-semiotic fields of meaning' (1991, p. 195).

Obviously, Haraway (1997) does not use location here as 'the concrete to the abstract of decontextualization' (p. 37). As she points out, we continue to rest on a whole metaphysics of essences, rather than take objectifications as provisional albeit 'stabilized interactions in a given frame of reference'. On the contrary, for Haraway,

> Location is the always partial, always finite, always fraught play of foreground and background, text and context, that constitutes critical inquiry. Above all, location is not self-evident or transparent. ... No layer of the onion of practice that is technoscience is outside the reach of technologies and critical interpretation and critical enquiry about positioning and location. (p. 37)

Haraway (1991) contends that, in order to go beyond the simple deconstruction of scientific objectivity pursued by today's scientists, we need to bring epistemological debate into the political and ethical fields to account for specific histories and engage in critical practices at the same time.

This project of an embodied and embedded situated objectivity, equally far from the nowhere of universal totalisations and the everywhere of relativism (and their common denial of an investment in location), finds in partial perspectives 'from somewhere' (p. 196) the context in which knowledge meets responsibility; and, here, responsibility means being locatable, and being able to give account of one's own locatedness (p. 191). Only when situatedness comprises the double gesture of accountability and responsibility does 'relativism ... redefined as partiality ... [become] an epistemic device' (García Selgas, 2004, p. 306).

Whereas postcolonial and transcultural feminist interpretations of the politics of location have usually expanded on the issue of difference in combination with geographical and historical questions (Mani and Frankenberg, 1993; Kaplan, 1994; Mohanty, 1995; Brah, 1996; Friedman, 2000; Anthias, 2002; Carrillo Rowe, 2005), the scholarship of feminist technoscience has privileged issues concerning materiality in relation with difference, particularly in light of the developments of situated knowledge and standpoint epistemology (Braidotti, 2007). This has paralleled a return to materialism in feminist theory that has tried to overcome the dichotomy between language and reality, focussing on the complexities of the 'material-discursive' which Haraway had already extensively foregrounded (Kember, 2003; Alaimo and Hekman, 2008, p. 6). As a result, many hybridisations between these lines of thought, as well as several starting points for cross-cultural work, have been advocated. For instance, whereas Massey (1992; 2005) relies on physics to restate that there is no absolute spatial dimension in which interrelations between subjects/objects take place, but rather a space-time complexity that gets constituted through interactions, Mei-Po Kwan (2004) proposes a hybrid geography that not only negotiates hybridity as a 'location' amongst geographical fields inside the discipline, but also overcomes the divide 'between the social-cultural and the spatial-analytical, the qualitative and the quantitative, the critical and the technical, and the social-scientific and the arts-and-humanities' (p. 760).

Accordingly, Sarah Whatmore (2006) notes that a '*re*turn' to materiality joins with the interest for 'vital connections between the geo (earth) and the bio (life)' in cultural geography and the polemics against representational methodologies in science and technology studies. This is associated, in Whatmore's view, 'with the intensification of the interface between "life" and "informatic" science and politics' (p.601), or the way the *livingness of matter* is redistributed amongst different actors and is also always in-the-making. Materiality does not stand for 'the indifferent stuff of a world "out there", articulated through notions of "land", "nature" or "environment"' anymore, but regards 'the intimate fabric of corporeality that includes and redistributes the "in here" of human being' (p. 602).

The idea of return that Whatmore suggests eschews the usual recourse to the novelty of the nth-turn by underlining instead the variations of continuity inside feminist thought. Drawing on Whatmore's notion, Hughes and Lury (2013) have recently examined the *patterns*, a term which encompasses continuity and differentiation, of feminist methodologies, focussing on re-turns as the 'turnings over' (p. 787) of situated knowledge epistemologies in the context of today's ecological epistemologies. While some feminist concerns, such as the insistence on situatedness, relationality, and the link between epistemology, ethics and politics, continue to be paramount, others such as the performativity of knowledge practices and the distribution of agency in more-than-human relationships, although certainly not new for feminism, become heightened in the 'expanded habitats' that characterise an 'ecological approach' to situatedness (p. 788). Feminist thinkers are confronted and re-read through one another, like in the case of Haraway through her harawaian reading of Barad (2007), and concepts such as positioning, negotiation and partiality are themselves situated over and over.

2.2   Representations of a Different Kind

Ecologically considered, situatedness reconfigures creatively as 'a co-invention that, fractally, recursively, opens onto other co-inventions. … Emergent in the diverse processes of differentiation, the patterns of movement, that constitute the moving surface or ground of figures of knowledge' (Hughes and Lury, 2013, p. 795). The ecological approach which Hughes and Lury (2013) refer to benefits from an increased feminist engagement with science and technology studies and the consideration of the entanglements of sociotechnical assemblages inside technospaces (Hayles, 1999; Whatmore, 2002; 2006; Kember, 2003; Barad, 2007; Suchman, 2007).

If scientific knowledge most often 'loses track of its mediations', Haraway (1991) on the contrary argues that it should take into account 'the earth-wide network of connections' through which knowledges are *translated* (p. 187; see also Latour, 1994) and elaborate 'a different *kind* of theory of mediations' based on *a change in the metaphor* that regulates the relations between bodies and languages (pp. 174, 188).

Analogously, Barad (2007) draws upon the assumptions that location is configured as a form of 'specific connectivity' and that our knowledge 'entails specific practices through which the world is differentially articulated and accounted for' (p. 149).[2] Nonetheless, she discards what she considers to be a representational approach to mediation, one that correlates, from the outside, two separate, distinct entities, or the representation with the represented, as if there were an external mediator operating on a homogeneous, static realm of things (pp. 374–5).[3]

Taking Judith Butler's (1990) theory of performativity further, Barad in fact rejects representationalism altogether. Butler defines performativity as an enactment without interiority or anteriority, focussing on the signifying practices of gender constitution that are inscribed onto bodies; Barad (2007) presents a processual, performative account of matter, or 'mattering', as the 'ontological performance its ongoing articulation' (p. 149), in which the material is always already the material-discursive (p. 153) and matter is what *enfolds* together with knowledge. Knowledge, in this respect, does not concern the apprehension of objective facts by external observers, but the enactment of specific configurations (p. 91). Concepts 'articulate' and 'give account for' the world through specific practices (p. 149). Conversely, matter is always '*what it means to matter*' (p. 153), never a passive surface but an actualisation of meanings. The separability, or what Barad also calls the 'cut', between the subjects and the objects of knowledge is thus the enactment, rather than the precondition, of the world's performances happening in *intra-actions*.

If 'representationalism takes the notion of separation as foundational' (Barad, 2007, p. 137), talking of intra-action means considering, on the contrary, the '*mutual constitution of entangled agencies*' (p. 33) which do not precede but rather emerge through their intra-acting. Whereas conventional epistemologies have conceptualised science as a 'set of representations of reality', intra-actionist approaches consider science as intrinsically technological and performed through different practices, interpretations and applications (Harding, 2008, pp. 186–7; Suchman, 2012). Scientific knowledge cannot accurately represent the world from a distance, let alone its objectivity, but can only show how the world effectively works and how representations can adequately fit such workings (Latour, 1987; Haraway, 1997).

---

2   They form the basis for her interpretation of quantum physics and defence of a *nonrelativist, realist* position – which she intends to distinguish from the *nonrelativist antirealist* position of standpoint epistemology and situated knowledge (p. 44; see also Eglash, 2011), although it is possible to argue that what properly distinguishes Barad's position is the function that she attributes to *mediation* rather than the way she defines reality, as explained above. Incidentally, a strenuous defence of realism according to a radical constructivist position, for which objects and facts are made up but no less real for this – contra social constructivist approaches – traverses non-representational debates (see Anderson and Harrison, 2010).

3   It must be noted that, in fact, representationalism is usually coupled with an instrumental thinking of technologies (Bolt, 2004, p. 8).

In Barad's philosophy of agential realism, intra-action, as the practice of boundary-making that works ontologically as well as epistemologically, is an antidote to representational mediation (for Barad a synonym of *interaction*) as long as representationalism entrenches mediation in reflexivity, leaving the ontological and epistemological a priori distinction of subjects and objects unquestioned. For Barad (2007), 'reflexivity is based on the belief that practices of representing have no effect on the objects of investigation and that we have a kind of access to representations that we don't have to objects themselves' (p. 87). Representation, she continues, even when 'raised to the nth power', 'does nothing more than mirror mirroring' (p. 88). Paraphrasing Haraway, Barad affirms that, although '"seeing" takes a good deal of practice', the disjunction of what we take to be the evidence and the practices that have produced that evidence are what make representationalism ignore 'the significance of practices' of vision (p. 53). However, as will soon become clear, Barad does not bracket optics altogether, but rather retrieves Haraway's (1997) notion of diffraction as the process which, instead of multiplying the effects of reflection, eventually disengages knowledge and representationalism.

Notwithstanding Barad's dismissal of it, the performative role that mediation plays in Haraway's (1991) theory is, in the end, very similar to the notion of intra-action. As a matter of fact, Haraway delinks representations from representationalism and, upon this, she builds a different theory of mediation[4] which is deeply linked with the politics of situated knowledge and optics. Haraway forcefully makes explicit what – if anything – remains ambiguous in the politics of location and in the notion of standpoint: that there are no privileged or innocent positions, and that every position must be interrogated, learnt, assumed or revised through the exercise of a semiotic-material technology: '*How* to see from below requires at least as much skill with bodies and language, with the mediations of vision, as the "highest" techno-scientific visualizations' (p. 191). And, she crucially remarks: 'The moral is simple: only partial perspective promises objective vision' (p. 191).

When Haraway (1991) talks about 'the embodiment of all vision', she does not necessarily intend it as 'natural' embodiment, but as an embodiment which includes 'technological mediation' (p. 189). In fact, no unmediated vision exists. On the other hand, whereas both natural vision and technologically-aided vision can be made transparent and their mediation erased (p. 189), 'there are only highly specific visual possibilities' (p. 190) that, in turn, are generative of partial perspectives. That is why understanding how optics works, and how 'instruments of vision mediate standpoints', who has the power to see, and how this power is employed at different levels, is of the utmost importance for a feminist 'politics of positioning' (p. 193). For example, in *The Age of the World Target*, Rey Chow (2006) asks what politics of vision the dropping of the atomic bombs in Japan supports, and what representations are instead precluded. Chow believes that such an optics conflates the visibility and the objectivation of the world to

---

4   In this respect, Haraway's idea of mediation is very near to Latour's (2005).

the point that the world becomes a target which is destroyable as soon as it is made visible.

Haraway redefines a 'political semiotics of representation' in non-representational terms, using the concept of articulation (1992, p. 311) to foreground the performativity of a mediating practice in which both the representor and the represented stay on the same ground of action, where 'boundaries take place in provisional, neverfinished articulatory practices' (p. 313). The 'crucial boundary breakdowns' between human and animal, organism and machine and the physical and non-physical realms that have put an end to the 'border war' of Western science and politics today, involve the territories of production, reproduction and imagination (Haraway, 1991, pp. 151–3). Some important transdisciplinary shifts in scholarship, according to Whatmore (2006), have followed such breakdowns. The first shift is the relocation of agency in practice and performance, and a re-embodiment of theory itself, which marks the passage from discourse to (informed) practice. The second is the shift from meaning as the production of a disembodied consciousness to affect, involving a rediscovery of the precognitive and its role in sense-making as a 'force of intensive relationality' (p. 604). The third, a consequence of the previous dislocation, is the shift from the human to the more-than-human, or from society conceived as a closed and exclusively human whole to a multiplicity of assemblages constituting a heterogeneous fabric. Finally, the fourth shift is the move from a politics of identity to a politics of knowledge, produced, negotiated or contested in different sociotechnical environments and along distributed practices (pp. 603–4).

Thinking about the relationalities that emerge from such shifts, then, means overcoming the oppositions that counterpose absolute, self-contained entities against each other. If the existence of the master subject was, in fact, guaranteed by the possibility of a separation from the object – something which very often happened through a distancing, reflecting vision – an articulatory turn in representation as Haraway (1991) imagines it also means the end of the same possibility of this Subject's existence and its substitution with 'non-isomorphic subjects, agents, and territories' (pp. 192–3).

In the last two decades, the debate around the issue of representation has occupied several different fields, primarily as a reverberation of the anti-realist constructivist turn that has permeated postmodern philosophical debates. For example, discussing the different traditions of the conceptualisation of representation as the knowledge of reality, Markus F. Peschl and Alexander Riegler (1999) show that a change of focus has occurred, especially after contemporary theorisations of autopoietic systems and developments in neuroscience, shifting from an attempt to grasp the structure of reality and map it onto a representational structure, resulting in *referential* representation, to an awareness of *system-relative* representation as a non-transparent and generative process in which the environment, rather than reality, 'constrains' what becomes representable instead of determining its outcomes.

The acknowledgement of the boundary breakdowns and of multiple processes of mediation identified by Haraway takes the form of a strong critique of representation in non-representational theory in particular. This, in most cases, associates representation with the metaphysics of visualism, although to paraphrase Andrew Pickering (1994), when vision is delinked from 'the representational idiom' and rather aligned with the 'performative idiom'(see also Anderson and Harrison, 2010, p. 19), a recovery and redefinition of visuality always becomes possible. Furthermore, despite the prefix non-, non-representational theories do not refuse representation *in toto*, but bring attention to what, in re-presentations, re-turns as not the same, to the 'transformation and differentiation' processes that the repetitions of representation bring forth through different media, rather than obeying the 'duty of resemblance' of one medium to the other: non-representational theory, in fact, supplants reality-medium relations with medium-medium relations (Doel, 2010, p. 119).

The terms of the debate regarding non-representational theory were initially assessed in the field of human geography, but soon turned out to be of interest for many other theoretical domains, such as feminist studies, performance studies and science and technology studies (Lorimer, 2005). In non-representational theory, knowledge is firmly located in matter or, to partially paraphrase the subtitle of Barad's (2007) book, in 'the entanglements of matter and meaning'. It is also relationally generated, and in no way a solely rational, subjective or even human property, these all being assumptions that, on the contrary, belong to the tradition of Western Modernity (Thrift, 2008, p. 122).

As Nigel Thrift (2008, p. 113) shows, non-representational theory has its roots in different philosophical traditions and their reciprocal points of contact. To name but a few of these, they include: feminist theories of performance and feminist spatial analyses, ranging from Butler (1990) to Irigaray (1991); the theory of practice drawing on the work of authors such as Bourdieu (1997) and de Certeau (1984); and what goes under the name of 'biological philosophy', from Deleuze and Serres to the current speculations in biosciences (see Rabinow, 1996).

Thrift (2008; see also Anderson and Harrison, 2010) characterises non-representational theory as the conjoined insistence on a number of factors. It features a radical empiricism – which is anti-essentialist in character and distances itself from constructivism – while aligning itself with philosophies of becoming, without completely abandoning the lived immediacy of the phenomenological and the precognitive. It includes an anti-subjectivism that disengages perception from the human perceiver and attributes it to encounters amongst heterogeneous forms, or what he calls 'new matterings' (p. 22). It relies on practices as being generative of actions rather than being their consequences, thus showing an interest in the 'effectivity' of the world (p. 113). It insists on *life* as the more-than-human vital coimplication (Anderson and Harrison, 2010, p. 12) of bodies and things in a network of functions, where embodiment becomes a diffuse situation of shared relationality. It requires an experimental attitude, which owes much to the performing arts and is based on the unpredictability and radical possibility – as the possibility of change – of the evenmental (p. 114), intended as what does not pre-exist the contingency of its own taking place, and for this reason is the anti-representational *par excellence*. This is well explained, for example, in what Anderson and Harrison (2010) write about representations as *doings*: actually, 'even representations become understood as presentations; as things and events they enact worlds, rather than being simple go-betweens tasked with re-presenting some pre-existing order or force' (p. 14; see also Doel, 2010, p. 120). Moreover, non-representational theory takes an affective stance that allows the retention of a sort of 'minimal humanism' (Thrift, 2008, p. 13), while at the same time being anti-humanistic in a traditional sense, and which translates into an affirmative ethics of responsibility and care. Finally, it has a situational character in which space is becoming, distributed and networked.

Notwithstanding the assumption that doing without representation is impossible, and that a recovery of the sense of vision, or better, of revision, is of the utmost importance for the feminist project of a multidimensional cartography of the world, Haraway's (1991) theory of situated knowledge presents many similarities with, and to some extent anticipates, the premises of non-representational theory,[5] as much as, in turn, current feminist scholarship has further elaborated (on) its links to non-representational theory, trying to uncouple the critique of representationalism and a supposed refusal of all representations (see Jacobs and Nash, 2003). Haraway (1991), in fact, believes that a simple opposition to representation advanced in the name of the world of matter is risky, being implicated in the double bind that sees matter and meaning as standing in a relation of mutual exclusion. Instead, 'a map of tensions and resonances between the fixed ends of a charged dichotomy better represents the potent politics and epistemologies of embodied, therefore accountable, objectivity' (p. 194; see also Kember, 2003). For these reasons, she insists that we not stop to pose the following questions:

> How to see? Where to see from? What limits to vision? What to see for? Whom to see with? Who gets to have more than one point of view? Who gets blinkered? Who wears blinkers? Who interprets the visual field? What other sensory powers do we wish to cultivate besides vision? (p. 194)

As Jane Jacobs and Catherine Nash (2003) have affirmed, commenting on recent scholarship in cultural geography, there is no need to dismiss representation altogether, particularly if we consider the importance of a critique and a politics of representation, even when sharing the assumptions of non-representational theory. As they put it, if we 'insist on attending to the place of image', we can keep open a 'wider semiotic framework' in which words and things interrelate, without contradicting the semiotics of materiality of non-representational theory (p. 273).

It is in this direction that Hayles (1997) has looked for an escape from the alternative between realism and anti-realism through her notion of 'constrained

---

5   An evident common root is Latour's ANT (Latour, 2005).

constructivism', which does not tell us what reality is, but rather what fields of possibility make certain representations 'consistent' with reality, and thus practicable for us. What we call 'observables', she writes, always depend on locally situated perspectives according to which different pieces of information about the environment are processed, as demonstrated in the example of the frog's visuality which Hayles gives at the beginning of her essay, drawing on the well-known article of Jerome Y. Lettvin et al. (1959). For the frog, the Newtonian first law of motion, which for humans applies to every object upon which a force is exerted, does not work equally. A frog's brain is only stimulated by small objects in rapid movement, allowing it to detect potential prey, whereas bigger or static objects elicit a completely different response.

Recognising, however, that every reality is relative to the observer does not lead Hayles to conclude that systems close in upon themselves leaving the world outside, that reality is infinitely suspended or that perceptions can do without representations at all. As Hayles (1995) notes, even if we agree with the non-representational aspect of perception, we do not necessarily need to believe that 'it has no connection with the external world', particularly when we consider that a relation can also be transformative rather than solely reflexive (p. 75). Not willing to renounce a term like representation, but rather intending to formulate it differently, as 'a dynamic process rather than a static mirroring', Hayles follows Niklas Luhmann (1990) when he recognises 'that closure too has an outside it cannot see' (Hayles, 1995, p. 98). This leads us to acknowledge, on the one hand, the fact that 'the very interlocking assumptions used to achieve closure are themselves the result of historical contingencies and embedded contextualities' (p. 98). On the other hand, it allows for a preservation of the 'correlation' or 'interactivity' that partial connections, rather than absolute distinctions, make possible (Hayles et al., 1995, p. 16; see also Guattari, 1995). Representations, in this context, appear not as a mirroring of 'external' reality, but as 'species-specific, culturally determined and context-dependent' processes of dynamic interaction (Hayles, 1997).

Hayles's (1997) constrained constructivism is built upon an 'interactive, dynamic, locally situated model of representation', in which the notion of 'consistency' replaces that of 'congruence'. Whereas congruence implies a one-to-one correspondence between signs and things, based on Euclidean geometry, consistency eschews this oppositional logic; rather than being enclosed within the true/false dichotomy, it stands in between the not-true/not-false relation, which is one that subverts the symmetry between affirmation and negation. Representations are ruled by constraints, which do not tell us what reality 'in its positivity' is, but can tell us when representations are consistent with reality, enacting some possibilities and enabling certain distinctions instead of others. Constraints, then, operate in the making of selections between those representations which are viable (or consistent) and those which are not.

Consistent representations are neither false nor true: they are either not-false or not-true. If I, for instance, look at the pen lying on my desk, I can surely say that it is an orange pen. However, my assertion is based on the observation of the colour that the plastic case of my pen appears to be. But if someone asks whether I have a black pen to lend, I can surely give them the same pen, given that it writes in black ink; thus it is a black pen, too. While asserting that my pen writes in black ink, I am not negating the orangeness of my pen, so to speak, but only further specifying something about the way it works. Consistency, then, should be intended as a relation of articulation in which 'articulations emerge from particular people speaking at specific times and places, with all of the species-specific processing and culturally-conditioned expectations that implies' (Hayles, 1997).

That consistent representations are not true but also not false makes them inhabit a kind of 'elusive' negativity that is neither negative nor positive. Consistent representations cannot be, so to speak, 'appropriated' by either side of affirmation and negation; they rather take place 'at the dividing line' (Hayles et al., 1995, p. 34). As for Haraway's (1992) *inappropriate/d other*, such alterity is not the untouched, authentic other, or the absolutely different, but the other that stands in a 'critical, deconstructive relationality, in a *diffracting* rather than reflecting (ratio)nality' (p. 299). It is not that we only partially see the truth in things while remaining ignorant of their totality. It is, rather, that partiality is the whole that we see, precisely as the result of constrained interactions.

2.3   Mattering Light: A Diffractive Methodology

Constrained constructivism presupposes a language of metaphors: the difference that passes between metaphors and descriptions is, for Hayles (1997), the same that passes between consistency and congruence. Haraway (1997) prefers speaking of *figurations* to name such 'performative images that can be inhabited' (p. 11).[6] Even though figurations always retain a visual aspect, which is not a secondary element in our 'visually saturated technoscientific culture' (Haraway, 2000, pp. 102–3), figures need not be literally representational or mimetic but made to 'trouble identifications and certainties' (Haraway, 1997, p. 11): they are neither complete nor static pictures of the world, but are representationally adequate insofar as they keep their performativity, with all its contradictions, alive.

Figurations, as modes of both tracing past connections and imagining new ones, are ways of 'figuring together', that is, con-figurations. Suchman (2012), drawing on Haraway, has recently underlined this conceptual node, stating that 'figuration … is an action that holds the material and the semiotic together in ways that become naturalized over time, and in turn requires "unpacking" to recover

---

6   Rosi Braidotti (2003) explains that this distinction between figurations and metaphors is intended to overcome the classical dichotomy of identity and alterity. From a Deleuzian perspective, the figural, based on difference and becoming, is opposed to the traditional aesthetic category of the figurative (or traditional representation) which, on the contrary, is based on identification and analogy between sign and object (see also Braidotti, 2002; 2006).

its constituent elements'(p. 49). According to Braidotti (2006), figurations map the metamorphoses and hybridisations of subjectivities in technoculture. They do not stand outside the world they describe but are living and transformative accounts which are never detached from their locations; they serve to 'represent what the system had declared off-limits' without, in turn, attributing a separate status to it (p. 170), stressing transition, interconnectedness, interaction and border-crossing. Figurations are thus *trópoi*, in that they, according to Greek etymology, do not simply figure, but 'turn' what they figure; that is, they act as performative mediations (Haraway, 2008a, p. 159; see also Latour, 2005; Suchman, 2012). Since they transform an exterior relation of correspondence into a relation of coimplication (Haraway, 1997; Lury, Parisi and Terranova, 2012), they are of the utmost importance for a project of technoscience intended as a 'travelogue of distributed, heterogeneous, linked sociotechical circulations' (Haraway, 1997, p. 12).

Haraway (1997; 2000) traces the origin of the meaning of the practice of figuration back to the semiotics of Western Christian realism, on the one hand, and to Aristotelian rhetoric on the other. In the history of Catholicism, the literal and the figurative continuously intersect, and figures are attributed with the power to contain the development of events, either of salvation or damnation – something which Haraway also finds in the millenaristic tone of many discourses of technoscience.

Aristotle (2012) highlights the spatial character of figures of discourse: in his philosophy, 'a figure is geometrical and rhetorical; topics and tropes are both spatial concepts' (Haraway, 1997, p. 11). As we have seen, in feminist thought, knowledge, representation and location have always been deeply intertwined: 'the politics of knowledge is understood in terms of the politics of representation, and the politics of representation is interpreted in terms of geopolitics of location' (Rose, 1996, p. 57; see also Robinson, 2000). This spatial aspect is visible in the strong link that Haraway's figurations, in fact, maintain with location, although clearly locations cannot be made to coincide with measurable space. Instead, they outline a cartography of relations and make sense of the different positionalities that these define (Braidotti, 2003), what Haraway (1991) calls a 'geometrics of difference and contradiction' (p. 170). Equally importantly, however, figurations also retain a temporal aspect that assumes the modalities of 'condensation, fusion and implosion', which is contrary to the modalities of 'development, fulfilment and containment proper of figural realism' (Haraway, 1997, p. 12).

Figurations are delinked from the theology of representation that revolves around reflection and reflexivity and their root in the 'mastery' of light (Haraway, 1992; 1997; 2000). 'The photological tenets of western philosophy' establish a strong correlation between light and visual representation (Bolt, 2004, p. 128). Whereas it is usually light that *sheds light*, that is, that unveils or informs matter, as the expression used to explain such 'clarification' manifests so well, the fact that matter is always in a process of *mattering*, and is thus always informing matter, is seldom considered. Even in the opposition between empiricism and rationalism (or mysticism), that is, between the idea of light as perceived through the eyes, and light as an ideal or divine source of enlightenment, representations are either the reflections or the prototypes of objects, so that representationalism still rules (Bolt, 2004, p. 125 ff.). If on the one hand Haraway (1992) affirms 'I do not turn away from vision, but I do seek something other than enlightenment' (p. 296), on the other she also interrogates the phenomenon of diffraction to find a powerful figuration for *mattering light*, so to speak, so as to dismantle at the same time the representational hierarchy that gives light an ideal pre-eminence over matter.

As a joke, albeit a serious one, Haraway (2000) affirms that semiotics is a science of four branches, 'syntactics, semantics, pragmatics and diffraction' (p. 104). Intended as the production of difference patterns, diffraction, the fourth 'optical' branch of semiotics, treats light differently from reflection though, as we will see, not necessarily in opposition to representation. As Barad (2007) so poignantly summarises, 'first and foremost ... a diffractive methodology is a critical practice for making a difference in the world. It is a commitment to understanding which differences matter, how they matter, and for whom. It is a critical practice of engagement, not a distance-learning practice of reflecting from afar' (p. 90). If reflection and reflexivity have their roots in representationalism (p. 87), the opposite is not necessarily true.

In this respect, Barad's (2003) assumptions about the dynamism and articulation of matter, which is not 'a support, location, referent, or source of sustainability for discourse' or any other external force inscribing onto it, but 'always already an ongoing historicity' (p. 821), are not so different from Haraway's appeal to the embeddedness of figurations. I thus disagree with Kirsten Campbell (2004), who sees a presumed evolution regarding the issue of representation in Haraway's writings, because I think that the model of articulation that a practice like diffraction presupposes is analogous to the way representations are reworked according to the notion of figuration, a project already pursued by Haraway in such writings as 'Situated Knowledges' (Haraway, 1991). I would not counterpose the latter to texts like 'The Promises of Monsters' (1992) or *Modest Witness* (1997) in which, according to Campbell, Haraway seems to abandon the representational model in favour of the diffractive one. Rather, what Haraway has always disregarded is the *metaphysics* of representation, while at the same time articulating representations by means of either situated or diffractive practices so as to render them viable.

Haraway (1997) never abandons representations, nor opposes diffractions to them. If Barad (2003) thinks that we should leave representations behind decisively for 'matters of practices/doings/actions' (p. 802), Haraway (1991) restates that seeing, too, is a doing, and that we are responsible for the 'generativity' (p. 190) of our visual practices. Accordingly, when Barad (2007) discusses the functioning of scanning tunnelling microscopes, which allow not only the visualisation of atoms but their manipulation, she notes the 'condensations or traces of multiple practices of engagement' (p. 53), the representational practices that produce evidence; our belief in them depends on contextual variables, so that critically engaging with representations is always possible and also desirable (Haraway 1997; 2000; Barad, 2007).

Diffraction gives light back its history (Haraway, 2000, p. 103). In fact, diffraction is a physical phenomenon which records the patterns of difference caused by the movements of rays resulting from *the passage* of light through a prism or a screen: 'a diffraction pattern does not map where differences appear, but rather maps where the effects of difference appear' (Haraway, 1992, p. 300). This process replaces the idea of a mimetic mirroring proper of reflection and refraction, or what Haraway calls the displacement 'of the same elsewhere' (1997, p. 273) – usually employed as a metaphor for the objectivity of science as well as for the traditional notion of artistic representation – in order to encompass interference, difference and interaction instead. 'To make a difference in material-semiotic apparatuses', says Haraway (1997), we must be able 'to diffract the rays of technoscience so that we get more promising interference patterns on the recording films of our lives and bodies' (p. 16). The historicity of diffraction, then, lies in its situated, embodied character and in its involvement in facticity and process-making.

Literally speaking, diffraction concerns the world of physical optics rather than that of geometrical optics. It describes the behaviour of waves when they encounter an obstacle, thus, practically all optical phenomena; contrary to geometrical optics, it also interrogates the nature of light. In physics, as Barad (2007) explains in her analysis, diffraction experiments are frequently used to compare the behaviour of waves to that of particles. One way to observe the phenomenon of diffraction – which the naked eye can easily notice when a pebble is launched into water or in the iridescence of a soap bubble – is the two-slit experiment, in which diffraction patterns resulting in bright or dark spots on a target screen – depending on the reciprocal enhancement or destruction of waves – are obtained when a light source passes through a two-slit screen (p. 71 ff.).

According to classical physics, only waves can produce diffraction patterns because only waves, not particles, can simultaneously occupy the same place. Barad, however, explains that quantum physics studies how particles can also behave like waves under certain circumstances, and discusses the 'modified' two-slit experiment at length, drawing on Bohr's (1958) diagrams. Without entering into too much detail here, it suffices to say for the purpose of our argument that, depending on the apparatus used in the experiment, that is, whether a 'which path detector' is employed or not, matter, and light as well, are observed to manifest either particle or wave behaviour. This apparent paradox forces us to radically rethink the dualism that lies at the core of representationalism and the idea that 'practices of representing have no effect on the object of investigation' (Barad, 2007, p. 87), given that diffraction not only shows the entanglements of meaning and matter but is itself an entangled phenomenon.

This is why adopting a diffractive methodology as Barad (2007) does, drawing on Haraway's lesson, implies a profound rethinking of both Western ontology and epistemology (p. 83). It replaces the analogical methodology, which consists in relating two separate entities by way of an external observer, with a methodology that shows how '*practices of knowing are material engagements that participate in (re)configuring the world*' (p. 91). In fact, as Haraway (2000) writes, 'diffraction patterns are about a heterogeneous history, not originals' (p. 101), which is to say that a representation is not a sign that mirrors a separate external referent; rather, it is a diffractive practice that performs the co-emergence of both meaning and matter.

Agency is redefined as precisely 'a matter of intra-acting', from which the 'agential realism' at the core of Barad's (1999; 2003; 2007) philosophy is derived. Since 'intra-actions are *constraining* but not determinate' (2003, p. 826, emphasis added; see Hayles, 1997), intra-acting neither belongs to a completely free subjectivity nor to a fully determined reality, but rather happens in a hybrid environment where 'particular possibilities for acting exist at every moment, and these changing possibilities entail a responsibility to intervene in the world's becoming, to contest and rework what matters and what is excluded from mattering' (2003, pp. 826–7).

Going back to Haraway, her theorisation of diffraction very much complicates the notion of vision as well as that of location (and the situatedness of the observer), since it dismantles the exteriority upon which both have traditionally relied and replaces it with specific forms of connectivity as well as accountability. Even if the observer comes back, she does not stand in a separate domain, but is connected in continuous feedback loops with her cognitive processes since the closure of the observer's domain is never pregiven, but always achieved (Hayles, 1995, p. 78). As observers, we take part, writes Barad (2007), in the 'world's differential becoming' in which our knowledge enacts the world, engaging in 'specific worldly configurations' from the inside (p. 91). However, adopting a performative idiom as a substitution rather than as a diffraction for the representational one, and thus completely getting rid of representations, leaves a series of questions unresolved, as Hayles and Haraway particularly highlight. These concern the domain of the observer as much as the status of what is observable, but also that which relates the two sides of the intra-action (Barad, 2007).

Hayles's (1997) theory of constrained constructivism tries to formulate the viability of representations through the idea that they can never be congruent with reality but, rather, be consistent with it. Haraway's notions of figuration and diffraction serve to displace fixed identities and put boundaries in constructive tension, requiring engagement rather than distancing. While Barad (2007) recognises the importance of diffraction as a generative practice and interprets this notion in a non-representational way in her philosophy of agential realism, I have argued that there is no need to oppose diffractions to representations, since what Haraway abandons is, first and foremost, the metaphysics of representation, but not the performativity of figurations which can be read through and used to read through at the same time.

We configure our world and establish connections with it through our ways of seeing. Diffraction does not simply regard our visual field but is a practice that invests our knowledge, our imaginary and our actions at the same time. It is, as

Haraway (1997) writes, 'a ... technology for making consequential meanings' (p. 273). Productive interruption, as well as reciprocal reinforcement, is allowed by diffractions and their unpredictable and unintended effects: different realities and unforeseen possibilities can always emerge from diffractive practices (Haraway cited in Schneider, J. 2005, p. 150).

2.4   The Place of the Image

Although diffractive approaches have been fruitfully utilised in several different fields, from pedagogy to medical and disability studies (Gough, 1994; Elovaara and Mörtberg, 2007; Weinstein, 2008; Salmon and Bassett, 2009; Puig de la Bellacasa, 2012; Takeshita, 2012), and an entire issue of the journal *Parallax* (2014) has been recently dedicated to the notion of diffraction, I will focus here on the way diffraction has been employed in media and visual studies (Lynes, 2011; 2013; Kember and Zylinska, 2012; Smith, 2012), before using a diffractive methodology myself in this context.

Krista Geneviève Lynes has recently (2013) coined the term of 'prismatic media', drawing on the Harawaian optical figurations, to name those media interventions in contemporary culture that engage with the multiplicity and difference of positions while foregrounding mediation as an active process. In an essay (2011) that precedes *Prismatic Media* (2013), she discusses the expanded field of the visualisation of war and the diffusion of grassroots media, in consequence of which more actors are in the position of recording the visual experience of war. This provides 'more genuine', grounded perspectives, which contrast with pervasive official rhetoric taking part in a broader diversification of media, such as mobile media and social networks, that 'in themselves' are supposed to democratise the possibility of documenting what really happens. If the politics of visibility of official media is more easily identifiable as one founded on naturalised opposition between identity and alterity (the various phases of the US war on terror and their visual crutches), which often translates into a moral dichotomy between the positivity of visibility and the negativity of invisibility, the perspective from below, Lynes notes, does not necessarily correspond to an abandonment of the principle that certain representations can depict reality better than others and that representation works as far as it truthfully reflects what really happens. Thus, she asks: 'On the ground, yes, but whose ground? And on what grounds?' (p. 27).

Apart from employing many traditional representational conventions (such as the shaky camera or an abrupt disappearance of images to signal live authenticity), Lynes (2011) wonders whether such visualisations privilege certain ways of seeing at the expense of other ones: that is, whether they act as reflections that reinforce existing visual systems (and, conversely, certain epistemologies and politics), or instead, as Haraway puts it, as diffractions that interfere with identities and work towards retracing the effects of difference. As an example, Lynes examines a piece of footage shot by a Canadian soldier in Afghanistan with a camera mounted on his helmet which was aired on CBC. The video has many features analogous to the videos shot by embedded journalists as well as videogame aesthetics, accordingly equating invisibility with danger and the (Taliban) enemy. She then compares it with the mostly unedited footage shot with hidden cameras and posted on YouTube – because it is often refused by official channels – by the Revolutionary Association of the Women of Afghanistan (RAWA). The author notes how, in this case, the Afghan women activists consciously assume the invisible position which is usually imposed on them by the Taliban, while also having to negotiate their complex situatedness in the international field of feminist activism. As Lynes observes, acquiring or restoring visibility alone does not suffice if 'the terms of visibility are already predefined by a socio-cultural system and semiotic apparatus that only brings certain subjects into visibility, and only on specific terms' (p. 26). Instead, she continues, diffracting interferences must disturb such predefined systems in order to articulate the differences that are both inside and between visual cultures.

Kember and Zylinska's (2012) work is a four-handed, reciprocally diffracted analysis (Kember is an academic and writer and Zylinska is a visual artist) of media culture and the concept of mediation that proposes a performative critique of representationalism, which also keeps performativity distinct from constructionism, as in the tradition of material feminism. Although the authors do not directly refer to diffraction in their book, but rather to Hayles's notion of 'intermediation' (2005) – intended as the process through which bodies and texts, but also different media, entangle amongst themselves – and Barad's notion of 'intra-action', their conjoined reading of the 2007–2009 credit crunch and the 'big crunch' of the Large Hadron Collider (LHC) at CERN in Switzerland is an example of diffractive methodology. They take the particle collision of the LHC as a paradigmatic mediated event that, between the promise of recreating the conditions of the Big Bang and the threat of producing unintended back holes eventually swallowing the entire universe, has been so 'over-invested in scientific representationalism' as to be transformed into 'an object, or an autonomous thing, rather than a phenomenon or process incorporating multiple agencies' (p. 62).

The experiment of the particle collision, they argue, remains trapped in a representational rhetoric of 'causation' and 'discovery' (p. 32) that ends up erasing the 'multiagential' forces (p. 40) at work in its unfinished taking place. The scientific and televisual visibility of the particle collision, *represented* in photo galleries and animations, has developed around what could only be eventually seen, that is, the existence of the Higgs boson (aka the God particle) behind the 'smoking gun' traces of the particle collision, only barely visible on screen. What such a process brings to the fore, according to Kember and Zylisnka, is not so much the particle collision as the 'observed' event but rather the productive, generative forces of representing practices that should make us consider 'to what extent … they [are] helping to bring about the event that they subsequently describe' (p. 63).

For the authors, the performativity of mediated practices in fact enacts different 'statements, beliefs, images, and stories' that take place either in discourse or in material practices (p. 102). But other images and stories which do not acquire the same representational status remain in the background. Amongst the less visible aspects of the particle collision, as Kember and Zylinska observe, there are, for example, the reinforcement of an ideology of pure knowledge as an end in itself, together with an ideology of 'conciliation' achieved transnationally between the USA (as participating in the construction of the collider) and the European Union, that in turn furthers the idea of a community of people equated to particles transcending national borders; moreover, there is also the testing of the Worldwide LHC Computing Grid, a distributed system of data storage required by the high volume of data produced, eventually usable for 'tracking and security, expressly designed to prevent the free flow of information' (p. 67).

Drawing on personal engagement and, of necessity, on a declared partial perspective, Laurel Smith (2012) examines the video *Dulce Convivencia/Sweet Gathering* (2005), directed by her husband Filoteo Gómez Martínez. The video, which documents the production of unrefined panela sugar in the Mixe community of Oaxaca, Mexico, belongs to the tradition of indigenous videos. Smith defines the latter an example of 'postcolonial technoscience', as it negotiates its hybridity amongst multiple instances of location and mobility (p. 330). She thus conducts a three-fold analysis of the video in which she considers the audience response at festival screenings, the geography of its local production and translocal circulation, and the perspective of Martínez himself. These three strands of her analysis are then reciprocally diffracted in order to multiply their possible significations and to let their – often contradictory – productivity emerge.

Underscoring the risks of a notion like hybridity, so often assimilated by the nation state to promote a depoliticised view of indigenous authenticity and deprive indigenous people of actual agency, Smith instead appeals to Haraway's theory of hybridity as a useful tool to declare a situated 'politics of visibility' which at the same time never 'lose[s] sight of technological mediation' (p. 331). Smith draws on Haraway's double analysis in 'The Promises of Monsters' (1992) of an exoticised picture of a protesting Kayapo Indian in traditional dress while using a videocamera that, published in *Discover* magazine as an illustration of the article 'Tech in the Jungle', works as an example of nature/technology dichotomisation. To this, Haraway juxtaposes a situated perspective that also tries to include a consideration of the multiple connections that make this representation happen, in order to include the agency of the Kayapo Indian in the picture as well. However, as Smith poignantly notes, in Haraway such readings are made to diffract each other, since none can offer a final truth of the image, but only partial perspectives are possible. In fact, whereas essentialised images usually belong to a conservative agenda, they are also sometimes reclaimed by the marginalised as a guarantee of visibility: 'and that's the whole point of diffraction. Not everyone yearns to know about the same things' (p. 333).

So, in the account that Smith offers, the audience's reception is diffracted with the author's ideas, and at the same time the author's position is diffracted through a complex web of intellectual and academic collaborations, state sponsorships, media technologies and practices of local activism that mediate the video location and substitute a binary narrative of indigenous vs. non-indigenous perspective with more nuanced, articulated visions. In sum, Smith's intent is 'approaching indigenous videos as technology-mediated co-productions of indigeneity that arise out of contingencies, personal connections, and specific organizational geographies' (p. 339).

The analysis of the short film *Mirror Image* (2013), by Danielle Schwartz, to which the last section of this chapter is dedicated, is perhaps the most poignant example which can help us to understand how a diffractive methodology works at tracing connections that, without reaching a final closure on a univocal truth, nonetheless offer consistent accounts (Hayles, 1997) that occur when contradictory positions are articulated and mutually diffracted as opposed to being simply represented.

Schwartz is an Israeli filmmaker from Tel Aviv trained in literature and cultural studies and currently living in the USA, who works on Israeli visual discourse and the cinematic representation of Palestinian space in Israeli cinema. *Mirror Image*, her first film, has been screened at various festivals in the USA and Europe and was the recipient of the Van Leer Award for Best Short Documentary Film at the Jerusalem film festival in 2014. Conceived as part of a more complex project revolving around her family, the film takes its cue from the story of a large crystal mirror of Palestinian origin that stands in Schwartz's grandparents' house. 'I was working on it for about 8 years', recounts Schwartz, 'during which I was mostly asking myself what is the right visual way to use the mirror, since the mirror was the starting point for me and the visual and conceptual heart of the film' (Schwartz, 2014). Preceded by some interviews with her grandparents on video and audio tape, the film itself was shot in one day to be presented at the end of November 2013 at The International Film Festival on Nakba and Return organised by Zochrot, an Israeli NGO that Schwartz was involved with at that time.

In fact, after reading some comments on Schwartz's Facebook page, her grandparents told her they did not want to be part of the project anymore. This denial, however, rather than preventing Schwartz from shooting the film, became what effectively animates its final version. As the artist explains, 'I realised that this conflict – the question where they drew the line – is the heart of the story. So I asked my grandparents to have that conversation (over what I can or can't say when I talk about the mirror in my film) on film. They eventually agreed, and that's how the film was created' (Schwartz, 2014). In a sense, the film that we now see is a film about the impossibility of shooting a film, which in turn is about the difficulty of telling a story of a displaced object, which, significantly, is a mirror.

At the start of *Mirror Image*, before crossing the threshold of the countryside house of Schwartz's grandparents in Binjamina, Israel, the camera pauses on some images of the garden and main entrance, where a row of Israeli flags flutters as a garland over the door. While Schwartz's voice begins to tell the story of the place where her grandfather grew up, the camera rests on some shelves in the kitchen, where several carefully arranged boxes and bins store food, spices and other flavourings, not only communicating a sense of domesticity and intimacy, of a well-established everyday routine, but also of locality and spatial embeddedness that cooking habits very often convey.

After this short visual prologue, Schwartz accompanies the viewer inside the house, narrating about her grandfather, who grew up and used to live in Kfar Gibton, a small settlement between the Jewish colony of Rehovot and the Palestinian village of Zarnuqa. As soon as the grandmother, who is shown cooking, corrects her granddaughter about the wrong use of the term 'Palestinian' to designate what at that time was rather defined as an 'Arab' village, an editing cut shifts our attention to the grandfather, carefully wiping a big wall mirror adorned by an engraved wooden frame, on top of which we partially distinguish a menorah, the Jewish nine-branched candelabrum.

The mirror is positioned in a recess of the kitchen, the place where the conversation at the centre of the film takes place. Here, as we learn, the mirror hangs on a special wall purportedly built for it, in front of the doorway, as a familiar object that the grandfather received as a present from his father, but that presumably comes from a Palestinian house from which it was 'taken' during the *Nakba*. This term, meaning 'catastrophe' in Arabic, was first used by the Syrian writer Constantine Zureiq to name the displacement and forced expulsion of hundreds of thousands of Palestinians and the destruction of their towns and villages by Israeli forces before the Israeli State was declared in 1948.

Figure 2.1   Still from *Mirror Image*, directed by Danielle Schwartz, cinematography by Emmanuelle Mayer, 2013

Source: Courtesy of Danielle Schwartz.

When the camera zooms in on the grandfather's face, which we also see reflected in the mirror while he is still in the act of wiping it, his wife appears out of focus in the mirrored background, on the opposite side of the room, and in the meantime the voice of Schwartz goes on, telling about her grandfather's past, of when, as a child, he used to cross the fence between the Jewish settlement where he lived and the Arab village nearby. While the grandfather is speaking, mumbling in slightly different words what Schwartz is saying, we see Schwartz's hand modifying some terms on the sheet where the video script of *Mirror Image* is printed.

In this sequence, the surface of the mirror where the grandparents' and the granddaughter's voices meet and superimpose gives visual consistency to the several interferences that Schwartz stages in the course of the film. A variety of perspectives, experiences and identities not only converge in but are also *mediated* by the mirror: the grandfather's present perspective as well as his memories, the grandmother's present perspective as well as her memory; history as it has been officially narrated and lived personal experiences, those who retell and remember it and those who imagine and figure it; the affective, internal position of the artist as a member of the family and, at the same time, her negotiated belonging to the Israeli people and her institutional perspective as filmmaker (later, the grandfather will note that what really hurts about Danielle's project is that they perceive the story of their life being told in a too detached, 'dry' and mechanical perspective, which Schwartz consciously plays with, so as to constantly avoid the impression of a 'spontaneous' voice).

In fact, the mirror evokes more than one context, signifying multiple divides as well as encounters inside different contested spatio-temporal fields, such as the generational, the ethnic and the territorial. In this respect, the mirror works like the safety pin that Haraway (2000) talks about in her dialogue with Thyrza Nichols Goodeve. Here, Haraway (pp. 104–5) recounts that an undergraduate student that she had in Santa Cruz, who was also a midwife involved in the home birth movement, used to wear a safety pin on her hat as a symbol of her beliefs. While not negating the significance of the symbol in the context of its usage for the student, Haraway remembers that in her class she worked to put the pin back in the complex histories of its production and circulation, so as to reconstruct the chain of associations that the pin brought with itself without either privileging the student's perspective or dismissing other possible accounts. This return to visibility of an object's lost stories, however, did not channel the other meanings into one single representation, it rather served 'to make it impossible for the bottom line to be one single statement' (p. 105).

As Haraway explains further on in the same dialogue, objects work like 'frozen stories': no abstract knowledge in fact exists, since all knowledge of the world requires that its objects' stories are unpacked and enlivened, and no place in the world can 'take place' if not in history. In this respect, a diffractive methodology which produces interfering representations, needs to be based on the acknowledgement of the mediations through which not only a composition of forces (Latour, 1994) emerges in every practice involved in the mediating process, but also other spatio-temporal, apparently absent, actions can surface through their present *delegates*, which perpetually shift, that is diffract, actual meanings as soon as they appear.

Consider, for example, the installation *Unsettled* (2012–2013) by Cornelia Parker, a visual artist who, drawing on the tradition of Robert Smithson and Gordon Matta-Clark's environmental interventions, employs objects as traces of lost space-times. She represents these remnants in fragile, often dismembered, exploded or suspended assemblages, as if the lost stories behind the objects could be recalled by the trauma to which matter is subjected (see Willemse, 2010). In *Unsettled*, a series of cracked and broken wooden grids and planks lean against a wall. From a distance, they appear as useless debris and manifest the immobility of abandonment, but this is only a deception, since the nearer we get, the better we notice how they are suspended from the ceiling by a system of almost invisible wires. Interestingly, the wood in *Unsettled* has been collected by Parker on the streets of Jerusalem, an origin that immediately activates a series of spatio-temporal associations in the viewers, particularly when displacement is at issue. Parker's installation, thus, works like a trigger, a momentary condensation (see Haraway, 1997) in which the viewer's process of knowledge acquisition and the way the same object is viewed result in a heterogeneous assemblage of signs and things (see Latour, 1994).

'The point of spatial, temporal, and "actorial" shifting, which is basic to all fiction, is to make you move without your moving', writes Latour (1994, p. 39). Schwartz renders these shiftings not only by recourse to a displaced and displacing object like the mirror, but also, for example, through the reciprocal interruptions that frequently redirect the characters' dialogue, or by having Schwartz correct the script during the conversation that they all have around the kitchen table. Whereas the grandfather comments on Schwartz's words, saying that he did not only cross the fence to play with the animals, but also with the children of the village, though he says that there is no need to further fix the script, the grandmother interrupts him and says that, on the contrary, it must be said, because 'the kids were in contact with each other', and 'it *does* matter'. Schwartz, then, as if to retrace her grandparents' steps in words, repeats the lines of the script out loud, this time including what both grandparents have said and also adding a new piece of information about the fact that the house he used to visit was the village head's house.

After an emotional moment of friction between the grandfather and Danielle, in which he is irritated by her constrained condescendence, she asks if it is true that the mirror appeared in the house after the war. While her grandmother's face visibly stiffens as the granddaughter pressures them, the grandfather appears hesitant, as he seems not to remember the exact moment when the mirror appeared in their house. Then the shot zooms out to include Danielle in the frame, who sits on the opposite side of the table, right in the moment in which she, assuming a more detached tone, clarifies what happened to the Palestinian village of Zarnuqa during May 1948, when it was invaded by a Givati Brigade in the course of the so called 'Operation Baraq'. The state of tension increases, and this time it is the grandfather who interrupts Danielle, asking what this has to do with what they are talking about. She responds that she wants to tell what happened in Zarnuqa, to which the grandmother comments, as if she wanted to legitimate an account that she cannot properly articulate, that all this can be found in 'history books', that there is no need to 'worry about it'.

Although Danielle notices that they are bothered by her insistence, she still wants to give a context to this narration of the 'taken' mirror, and she thinks this is a way to do it. In fact, what Schwartz is trying to create here is what Haraway talks about in terms of the interplay of 'foreground and background, text and context, that constitutes critical inquiry' (1997, p. 37). No position remains the same from the beginning to the end of the film, but the encounters between them create dissonances and resonances that reciprocally diffract and continuously alter their initial tendency. So, for example, the moment when the grandmother tries to use a softer tone, saying that she understands the reason for these annotations, the grandfather keeps on saying that he does not see the necessity of delving into this so much. The fact is, Danielle repeats, that since the mirror has been taken from a Palestinian home, which is deducible from what they have told her about the object's origin from Zarnuqa, she supposes (this time when she talks, the camera focusses on a close-up of her face, so as to visibly focus on her point of view) that it could have been 'plundered' somehow: 'because taken during wartime equates plundered'.

But here again, the facts appear not so easily retraceable: the grandfather repeats that he received the mirror from his father, who told him he had bought it, but also 'taken' it in Zarnuqa. However, since he 'knows' his father, he also finds this account implausible enough, as it seems impossible to him that his father could have taken something improperly. This remains an unverifiable assertion, of course, but it nonetheless produces a partial account of what could have actually happened. So, the persistent circulation of the terms around which the story of the mirror revolves, the way these terms are either repeated or evaded by the participants in the conversation, signals that what cannot be represented can still be performed.

As Butler (1997) writes about the mechanisms of censorship, there is a paradoxical productivity in the constraints of a text, which eventually culminates in a proliferation of the omitted terms within the boundaries of regulated discourse: 'language that is compelled to repeat what it seeks to constrain invariably reproduces and restages the very speech that it seeks to shut down. In this way, speech exceeds the censor by which it is constrained' (p. 129). Moreover, the productive role of censorship also acts on the formation of subjectivities. In fact, 'the question is not what it is I will be able to say, but what will constitute the domain of the sayable within which I begin to speak at all', says Butler (p. 133). Once pronounced and circulated, even by refusal, these words locate the subjects of speech in a context that exceeds the representational field, and rather concerns the embodied position of the speaking subjects that are interpellated and in turn work as substitutes for the absent ones. The context in which the speaking/viewing subjects are represented, however, being performed rather than represented, never stands still, but 'is itself subject to a further contextualization' since 'contexts are not given in unitary forms. This does not mean, and never meant, that one should cease any effort to delineate a context; it means only that any such delineation is subject to a potentially infinite revision' (p. 148).

As we have seen, the mirror can be made to work as a reflecting or a diffracting plane, depending on the approach to representation that it enacts (Foucault, 1986; Coyne, 1999; Massey, 2005), whether it is one of distancing and identification or one of involvement and articulation. As a matter of fact, the material object whose story is, although incompletely, retraced, and the words that narrate this story, whose difficult, almost impossible utterability is nonetheless represented and given visual space on the cinematic screen, are displaced through the mirror in *Mirror Image*.

In this double movement, the void that the mirror and the words disclose is the space where reality and representation appear as simultaneously performed, in which Schwartz simultaneously occupies the position of the filmed narrator and of the filming cinematographer. In fact, her declared positionality, as both an internal and external observer of what she herself contributes to performing, foregrounds the process through which knowledge, what is known, and the knowing subjects are constituted together in a partial and relational enactment that is materially arranged and distributed across several boundaries. Such enactment is ultimately an 'ongoing open-ended articulation' (Barad, 2007, p. 379) that does not concern the intellectual subject alone, but requires a 'differential responsiveness (as performatively articulated and accountable) to what matters' (p. 380). It also puts into (inter)play both the text and the context of representation, entangling the discursive and the material plane.

Belonging to a non-unitary but multidimensional knowing subject, vision has a stratified topography that constitutes its limit but also its openness. So, like in the activity of vision, the lines that can be drawn between Schwartz's and her grandparents' words, between the mirror and the other objects in the house, and between Palestinian and Israeli spaces, are 'always constructed and stitched together imperfectly' (Haraway, 1991, p. 193). In this respect, Schwartz's intent is not simply to search for and document where truth belongs, but rather what cuts and junctions produce *which* and *whose* truth. Putting differences in tension between the time-space of representation and the representation of the time-spaces that belong to the her and her grandparents (also recalling other time-spaces of absent subjects which can only be evoked), she therefore becomes, and puts the viewer in the position of becoming, 'able to join with another, to see together without claiming to be another', as Haraway would put it (1991, p. 193).

As in the methodology of diffraction (Haraway, 1997), understanding how, where and for whom differences resonate is something that cannot be made from a safe distance, but requires that the actors involved, the tools and the approach adopted in the process be clearly situated. Actually, as Barad underlines,

> the point is not simply to put the observer or knower back *in* the world (as if the world were a container and we needed merely to acknowledge our situatedness

in it) but to understand and take account of the fact that we too are part of the world's differential becoming. And furthermore, the point is not merely that knowledge practices have material consequences but that *practices of knowing are specific material engagements that participate in (re)configuring the world*. Which practices we enact matter – in both senses of the word. Making knowledge is not simply about making facts but about making worlds, or rather, it is about making specific worldly configurations. (2007, p. 91)

So, as Haraway first writes (1991; 1997) and Barad repeats (2007), every optics presupposes but also produces a positioning, since when one's situatedness is accounted for, then the articulation of representation inevitably implies a politics and ethics of representation as well (see also Thiele, 2014). 'The "eyes" made available in modern technological sciences', writes Haraway, 'shatter any idea of passive vision; these prosthetic devices show us that all eyes, including our own organic ones, are active perceptual systems, building in translations and specific *ways* of seeing, that is, ways of life' (p. 190). Acknowledging the partiality and embodiment of vision allows accounting for the activity of visual practices and their producing forms of embodied and partial objectivity 'that initiates, rather than closes off, the problem of responsibility for the generativity of all visual practices' (p. 190).

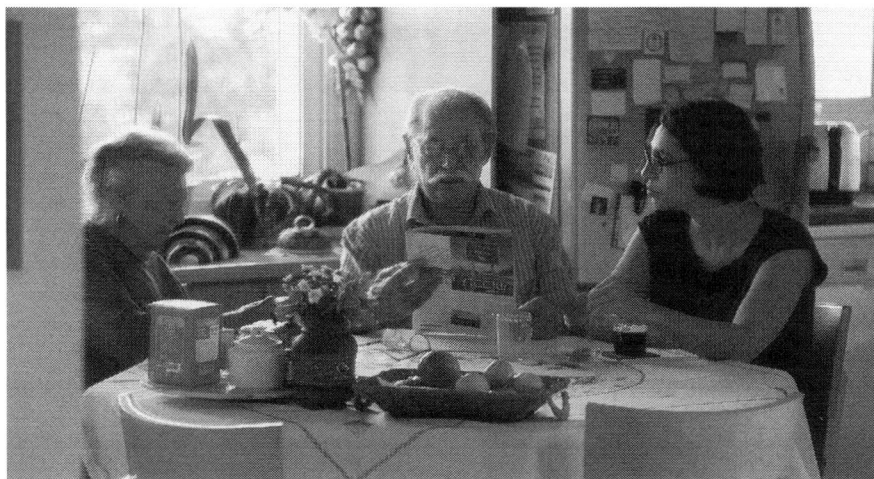

Figure 2.2   Still from *Mirror Image*, directed by Danielle Schwartz, cinematography by Emmanuelle Mayer, 2013

Source: Courtesy of Danielle Schwartz.

In *Mirror Image*, if Danielle's grandfather's words are diffracted by his granddaughter's annotations, the grandfather too has been diffracting his father's account over the years. Further, during the conversation, his account is in turn received and expanded by the grandmother, who affirms: 'we have it because that is what happened *to the mirror*, not *us*'. In this case, the attribution of agency to the material object, effectively acting as a delegate for the absent actors in the scene (Latour, 1994), seems to be able to relieve them both from the possible bad actions of their relatives. This is an account so rooted in the history of the family that, even when Danielle tells her grandparents to assume that there could also be proof about such plundering, her grandfather answers that he would never admit the occurrence of the event in any case.

Such an impasse is resolved when, paradoxically, the contradictions implied in private histories and intimate memories are bypassed by recourse to an external and apparently homogeneous temporality, that of wartime. Like before, when Danielle's grandmother appealed to history books to displace her bothersome personal memories in favour of their official counterparts, a separation is enacted again, this time between the family members and the 'people who are not like us', without needing to justify such people, or even forgiving them. Now the opposition, while clearly suggesting the us/them dichotomy of the Israeli/Palestinian conflict around which the mirror interface gains its more immediate significance, is brought inside the Israeli people, which further complicates the possible positions in the story. Again, this shift offers Danielle a loophole to expose how this alternate view of belonging and distantiation is, in the end, an escape from the assumption of a position of responsibility, which according to the politics of situated knowledges and the methodology of diffraction means being accountable not only for the material consequences of what we get to know, but for how the world is configured through the knowledges that we, accordingly, choose to perform (see Barad, 2007).

So, Danielle wonders why they feel so disturbed by the harshness of a term like 'plundering' if it does not seem to regard them directly. In what is perhaps the keystone of the whole discussion, the grandmother says that the problem basically lies in their visibility on the scene: 'we are in the picture', she notes, and this is why we would look as if we were '*like that*'. They, in fact, *are* 'like that', Danielle asserts: even if they are not thieves themselves, they have more than one thing in common with the perpetrators of these usurpations. They are Israeli people, and Israeli people have occupied and plundered a territory which belonged to 'the Arab', and so they too are, more or less indirectly, part of the ongoing Israeli-Palestinian conflict in some way. So, Danielle asks, 'what does make a difference?' If the things narrated are accurate enough, what does create this feeling of discomfort amongst them?

Mirror Image represents the difficult and incomplete process of knowledge acquisition through a diffractive visual register that investigates how vision is also always a question of power, the power to see which can disturb and disrupt existing visions. In this respect, it presents many similarities to the film *The Look of Silence* (2014), by the US director Joshua Oppenheimer. This is the second part of a diptych, whose first part is *The Act of Killing* (2012), in which the director explores the long forgotten history of Indonesian mass murders perpetrated against supposed communist sympathisers during the 1960s, after the Suharto coup in 1965. In *The Act of Killing*, the perspective explored is that of the perpetrators. The director makes their point of view explode through the artificialisation of their accounts, that they themselves perform as the 'actors' (this time in the sense of fictional characters) of stereotyped action movie sequences. The contrast between historical evidence and its mimicking in the *mise-en-scène* stresses the difficulty of accounting for the facts as they had actually happened for the spectators and the director, but also for the protagonists before them; it also unmasks the processes through which many events were officially kept secret to the point of making it impossible to uncover them if not through performing imaginative exercises, not less efficacious, in the end, than any objective report. *The Look of Silence*, on the contrary, assumes the perspective of one of the victims' relatives, Adi Rukun (as himself in the film), who retraces the story of his long lost brother, slaughtered before he was even born during the same political genocide recounted in *The Act of Killing*.

What is interesting for the comparison with Schwartz's *Mirror Image* is that, incidentally, Adi works as an optometrist who, when he goes around his village district to test his patients' vision, also urges them to narrate their memories. Whereas, obviously, Adi's vision-testing instruments are not simply the tools that he uses to calculate the 'refractive errors' of his patients, they can neither be intended as exclusively symbolic instruments that, all of a sudden, make the person wearing them unveil what really happened. As in *Mirror Image*, there is no quest for a final truth here, and no absolute keeper of the truth of history: truth instead emerges along a slow and uneven process in which Adi confronts many people in different contexts, collecting several pieces of words, pauses and omissions, which are then re-presented by the director in the place of the missing images.

In between his wanderings, Adi is also shown at home, while watching a TV monitor where he sees actual footage of accounts reported by the culprits of the crimes committed. Very often, we only hear what Adi sees, or see these images in fragments, while the camera instead focuses on Adi's face and his astonished, silent eyes. We watch a wordless Adi finally understanding 'by seeing', because since he wasn't there, he wasn't born yet and couldn't witness otherwise, this is the only way he can access the evidence of what happened. Yet, not only does he not have the words to verbalise the atrocities he listens to, he actually does not see anything except the perpetrators' descriptions. But the look of the silence on his face nevertheless *represents* the guilty silence that he tries to give voice to during his never-ending quest, where the truths of the victims, although ceaselessly deferred and incomplete, anyway resurface in the displacements between words and images, which are never fully overlappable.

As Oppenheimer (2014) affirms, his intent in this film is to immerse the viewers 'in the haunted silence in which survivors must build a life, surrounded by the still powerful perpetrators'. The only possible way of representing this is by creating a suspended space in which 'to pause, experience that and listen, and look, very closely to the silence that follows atrocities, particularly when there is no justice'.

When vision is deficient, bodies are fragile and words are lacking, silence assumes a disturbing 'look', one in which the view of the torturers and their more or less direct accomplices overlaps with the perspective of the victims, creating more than one interference. Like Adi's patients, Adi's father, who is still alive and maybe in his 100s, suffers from an eye pathology, to the point that he has almost completely lost his sight. He, who for his age and lived experience is supposed to be the guardian of the family's story, appears to be the least reliable amongst the family members. The mother, who is also very old, is shown taking care of him, carefully cleaning his body and occasionally washing his eyes in the hope of clearing his view. Although this can never actually happen there are moments in which the father seems to see better than the others, and is also able to answer the questions that Adi and his wife pose to him. However, this depends on the way the relation between Adi and his parents 'supports' the knowing process, which like in *Mirror Image*, proceeds through tears and sutures and always appears as a shared process.

When proper witnessing is impossible because of the many missing pieces of the story, performing witnessing *as if* this were the way things went is already a form of accountability, and for this reason it is uncomfortable, although not explicitly upsetting. Thus, and this is how *Mirror Image* proceeds, just when a difference is made, another link is established. The thematic alternation between connection and separation is achieved at a formal level by a continuous oscillation between the more affective voice of Danielle as granddaughter and the-often pretended-assumption of an objective standpoint as in a traditional documentary. This ambiguity confounds and irritates the interlocutors, who are never sure whether they trust Danielle as a narrator, but provokes the viewers' position as well who, thanks to the porosity of fictional boundaries, are called into question as part of a representation whose (never conclusive) signification their vision contributes to achieve.

In the final part of *Mirror Image*, to dampen the friction of the dialogue and move towards a progressive estrangement from the emotional involvement of the central section, Danielle, frontally framed and with a detached attitude, starts reading the videoscript that she has been holding in her hands and checking all the time during the conversation. As if it were a palimpsest, she has annotated, amended and overwritten it by considering the observations of her grandparents, their doubts and fears but also their omissions, which in some cases she also decides to compensate for: like when she pretends to forget to replace the term 'Palestinian' with the term 'Arab', her apparent slip testifying that, even if all accounts are partial, there are still 'less false accounts' (Harding, 2004b, p. 260) that need telling more than others.

In the conclusion, after a cut zooming out on the kitchen and showing the three characters together, a second cut brings the viewer back to the surface of the mirror, where we now observe the same image of the kitchen, reflected and initially out of focus, in an angle which is similar to the initial shot in which the mirror appeared with the grandfather wiping it. The image becomes sharper as soon as Danielle starts giving some information about her grandparents that we are only now confronted with for the first time: she tells about her grandfather's enlistment in the army at the age of sixteen, his training by the Palmach, one of the three brigades of the underground Jewish army before the official birth of the Israeli Army, his having fought in the Negev Desert, and finally their marriage and foundation of the Bar'am kibbutz, near the Lebanon border. It was after leaving the kibbutz, when they moved to their new house, that they received the mirror as a present from the grandfather's father.

Even if at this point the viewers have all the elements to compose their version of the story they prefer to believe, Danielle remarks the end of her account by underlining, on the one hand, the actual location of the mirror, and keeping the story open again on the other. In fact, in the very closing of *Mirror Image*, she does not offer any 'clear explanation', except reporting what she has been trying to figure out over the years. In so doing, she articulates what remains unknown rather than what has been discovered. Therefore, the mirror works like a visual litotes that, not being able to reflect a lost image, becomes the place where this image is looked through and represented *in absentia*. As Jay Bolter and Diane Gromala (2003) argue:

> When we look in a mirror, we see ourselves, and we see the room behind and around us – that is, ourselves in context. … The most compelling interfaces will make the user aware of her contexts and, in the process, redefine the contexts in which she and the interface together operate.

It is *in place of* the image that the conversation between Danielle and her grandparents takes place, but it is also *in the place of* the image that it is diffracted and given back to the context which hosts the mirror's displacement. Whereas, properly speaking, the mirror does reflect the kitchen of Schwartz's grandparents' house, its surface actually diffracts what we see so that something else appears in the picture: and apparently 'it's all true', says the grandmother, because right in the location where the mirror's original truth is forever deferred, truth can be nonetheless performed as a specific spatio-temporal configuration.

Chapter 3
Location, Mobility, Perspectives

3.1   Technospaces in the Middle

In theoretical analyses as well as in the practices of digital media, ideas about generalised connectivity, freedom of mobility and spatial boundlessness as a consequence of the extensive diffusion of new technologies play a big role. The various geographical metaphors and analogies employed to describe such aspects (Adams, 1997; Taylor, 1997; Brown and Lauriel, 2005; Serfaty, 2005; Lemos 2008; Kitchin and Dodge, 2011), from the electronic frontier and information superhighway to network topology, typically work to reveal what digital space has either in common or in contrast with physical space; they rarely serve to cast a different light on the way the space of new information and communication technologies could mobilise our imagination of space in general.

As a matter of fact, spatial metaphors not only have a descriptive function, but also productive and ideological ones (Adams, 1997; Graham, 1998; Holmes, 2000; see also Harvey, 1993) since they construct shared worldviews and figure how technology should 'express' society or relate to it. Considering digital spaces, the model based on the grid (Gordon, 2007) has been progressively replaced by a metageography of locality and a new rhetoric of ubiquitous computing with the advent of Web 2.0 and the massive diffusion of location-based technologies. However, even though 'the network has transitioned from a distant container for everyday life to a location from which everyday life emerges', as Eric Gordon writes (2007, p. 15; see also Rogers, 2012), leaving this location 'unchecked', so to speak, works as another form of (mis)representation.

The coexistence of contradictory descriptions of globalisation, alternatively defined as the epoch of absolute space and the epoch in which space is annihilated by time, should already make us reflect on the problematic nature of spatial dichotomies (Massey, 2005, p. 90). Is the sense of a loss of orientation and the experience of dislocation perceived as a consequence of time-space compression (Harvey, 1989), for example, valid for everyone, or is it rather a universal assumption masking the power relations upon which it rests, made up of differences, usually inequalities, among diverse sociospatial positions (Featherstone, M. 1993; Brah, 1996; Kirby, 1996)? Gregory (1994), for example, notes that Harvey's well-known formula still manifests a modernist sensibility that transforms a specific geographical imagination into a global norm. Pairing time-space colonisation, which is based upon an expansion towards the outside, with time-space compression, though it highlights the collapse of spatio-temporal coordinates, also ignores the differences out of which spaces are made (pp.

414–15). Thus, the pretension of a universal geographical imagination becomes a way to conceal the locus of its own emergence (see Duncan, J.S. 1993; Massey, 1994; 2005; see also Meek, 2000).

Every representation exists in context, which is to say that every representation is a representation of a space-time (Massey, 2005, p. 27). Nonetheless, when space is associated with representation, this mostly happens because representation has commonly been thought of in traditional spatial terms, as a way of fixing and stabilising represented things into a (pre)given frame. However, if both representation and represented space are mobilised, representation can be reconceived as an activity that does not stand outside the thing represented but which is entangled with it from the moment of its emergence. Accordingly, the opening, even the dissolution, of so called 'traditional' places, can be differently understood. How could we explain these occurrences after having attributed no processuality to places at all, unless we resort to an external, transcendental force – usually identified with a nonspatial progressively unfolding, very often technological evolution? On the contrary, if, from the very beginning, we assume an *extroverted* (Massey, 1994, p. 155) and generative sense of place, a space made out of interactions and intersections stretching across presumably natural boundaries, we can also imagine a past of already existing, although differently articulated, connections.

An important consequence regards the theorisation of identities and the necessity of breaking the representational symmetry that correlates community and place, what William Mitchell (2000) defines as the decoupling of *civitas* and *urbs* in network society. Geographical places are not the source of identity, nor are communities mappable onto them. As new ICTs today bring to the fore, communities can also exist without belonging to the same place. As Manuel Castells (1996) has shown, network society concentrates and, at the same time, disperses its territorial components so that simultaneity does not necessarily coincide with contiguity anymore. Besides, even while belonging to a common territory, members of the same community can have different senses of place, whereas the celebration of the seamlessness of digital space very often conceals an erasure of its specificities and tensions (Meyrowitz, 1985; Young, 1990; Massey, 1994; Appadurai, 1996; Bhabha, 1999; Mohanty, 2003).

Before the massive diffusion of electronic media, the existence of doors and entrances sanctioned a set of 'rules of physical place' for the social. The increasing mobility and speed of information in electronic media, having loosened the constraints of physical place on media environments, also implies a redefinition of the idea of social interaction in technospaces (Meyrowitz, 1985). Experience is not merely linked to physical location, but to *situation*, a more complex but less binding form of positionality. Massumi (2014), for example, strictly relates situation to information, as opposed to 'conformation': whereas the latter implies the application of a rule that pre-exists its context of enaction, which also is already given, situation consists in a 'taking-form' or 'form-finding immanent to the situated action', which does not presuppose anything preceding such an 'arena of activity' (p. 44). In the contemporary media environment, boundaries, which still surely exist and are continuously recreated, are redefined by the participation in the events of information and communication in which mobile interfaces continuously mediate the networks of sociospatial relations. How people communicate as if they were in the same place when in fact they are not becomes of the utmost importance. Now, location *situates* the actors of contemporary technospaces in a network in which 'the radical visibility of located data creates the potentiality for users to experience meaningful nearness to things and people' (Gordon and de Souza e Silva, 2011, pp. 12–13). Both the possibilities of choice and the dynamics of inclusion and exclusion increase.

So, what new ICTs urge us to consider is not what spatial forms cease to exist and what new forms emerge in our globalised scenario, but what relations are co-formed together with different kinds of techno-spatial configurations:

> The argument here is simply that what is, or should be, at issue in accounts of modernity and globalisation (and indeed in the construction/conceptualisation of space in general) is not a kind of denuded spatial form in itself (distance; the degree of openness; the numbers of interconnections; proximity, etc. etc.), but the relational content of that spatial form and in particular the nature of the embedded power-relations. (Massey, 2005, p. 93)

In his historical survey of the prevailing approaches to the issue of space and information technologies, Stephen Graham (1998) distinguishes three theoretical perspectives that have prevailed over time: *substitution and transcendence, co-evolution* and *recombinant*. The perspective of *substitution and transcendence*, typical of the 1960–1990s but still echoing today (Kaplan, 2002; Graham, 2004a; 2004b), sees new technologies as liberating us from previous temporal and spatial constraints to the point of a complete erasure of distance and a dissolution of the old geographies of territory and the body, considered as properties of the local dimension. Such a fantasy is imbued with technological determinism, neoliberal triumphalism, and a certain cyberlibertarian mystique, combining both utopic approaches, from Marshall McLuhan's (1962; 1964) to that of the cybergurus (Negroponte, 1995; Barlow, 1996), and dystopic ones, among which Paul Virilio's (2004) is the most well-known. Implicit, here, is the idea that space can be reduced to a question of measurable distance and that cyberspace is, to paraphrase the *Wired Manifesto* (1996), an 'infinitely replenishable and extendible' alternative territoriality (cited in Graham, 1998, p. 171). An outright theology of cyberspace flourishes from such an 'anything-anywhere-anytime dream' (Graham, 2004a, p. 4; see also Coyne, 1999; Wertheim, 2000), which loses sight of the mediated and mediating aspects of technospaces, where digital media interface with sociospatial dynamics.

The second perspective Graham (1998) examines, the perspective *of co-evolution*, believes in a parallel production of geographical and electronic space at various levels of reciprocity and influence, without neglecting the complex social and cultural dynamics in the production of place and space:

'new information technologies, in short, actually resonate with, and are bound up in, the active construction of space and place, rather than making it somehow redundant' (p. 174). This is what Castells (1996) – to draw on one of the several subjects of Graham's analysis – calls 'the culture of real virtuality' (p. 373), in which geometries of incorporation and exclusion complicate the rather simplistic view of a progressive shrinkage of the world to the point of total mobility without barriers. Neither ICTs nor places develop neutrally; the 'variable geometry' of the information society depends on the 'differential location' (Castells, 1996, pp. 145, 147) of the power dynamics at stake in global networks (see also Sassen, 1998; 2002; Mitchell, W. 2000) in which 'location' loses its exclusively geographical character in order to encompass the complexity of network culture. The problem with this perspective is that it tends to ignore the always-already networked aspect of place (Massey, 1994), considering it a recent phenomenon mostly linked to the diffusion of new ICTs. In addition, it leaves unexplained the way 'networks and interdependencies between people, technologies and places interact with the situated aspects of action within those places', as Barry Brown and Eric Lauriel (2005, p. 21) contend in their critique of Castells's network sociology.

A third perspective, drawing on Actor-Network Theory (ANT; see Latour, 2005) and on Haraway's (1991) thought, conceives of new technologies in terms of *recombination*, taking the relational view of technologies and societies further and reinforcing the relational consideration of space and time. In this context, relationality is what links human and non-human actors (such as technological artefacts) in contingent and heterogeneous combinations along multiple networks. Relations are, from the beginning, both technological and social. Drawing on the notion of 'technosociality', initially used by Arturo Escobar et al. (1994) and Allucquére Rosanne Stone (1995) to point to the interconnection of nature and technology in our lives, the recombinant perspective considers technospaces as the contingent contexts traversed by and performing the recombinations of society and technology.

The attention to the mediating operations which are proper to technospaces sets aside the belief in a dualism between the reality of space and its representation. Both real and virtual places, in fact, are seen as consisting of a 'fragmented, divided and contested multiplicity of heterogeneous infrastructures and actor-networks' (Graham, 1998, p. 178) that are contingently constructed and unevenly working. In these contemporary technospaces, the *where*, mobilised by information, finally allows the disengagement of spatiality from a purely dimensional perspective. Networked mobility, as the possibility of producing and consuming information in movement, cannot be considered as the passage from one point to another in space anymore, but, as Rowan Wilken (2005) puts it, it 'prompts renewed consideration of the "where" of everyday places … in transit'. This goes hand in hand with a different *a-whereness*, to use Thrift's (2008, p. 166) expression, an environmental understanding of the pervasiveness of technologies in the interstices of our everyday life. In this perspective, media are not what stand between society and space, but the active environment where different social actors share either processes of in-formation or of 'mattering' (p. 166; see also Grusin, 2010).

In this respect, W.J.T. Mitchell (2008) writes that media should rather be addressed than understood, as McLuhan would put it, 'as if they were environments where images live, or personas and avatars that address us and can be addressed in turn' (p. 3). Media, he continues, are like 'ever-elastic middles that expand to include what look at first like their outer boundaries. The medium does not lie between sender and receiver; it includes and constitutes them' (p. 4). Mitchell's aim is to show how media should be addressed rather than simply understood in order to, on the one hand, let the double communicative and spatial meaning of the term 'addressing' emerge and, on the other, underline the reciprocity of media as environments and of society as a complex system. He intends to expand McLuhan's (1965) ecological idea of media and, at the same time, criticise Raymond Williams's (1974) dereification of media, which leads the latter to ignore the materiality of media in favour of social practice as the only locus where actions become meaningful. Instead, for Mitchell, media are both the objects and the operations, both the complex of (human and nonhuman) technologies and the habitat where they *take* and *make* place (p. 9). Media address us when they acquire their spatio-temporal location. Thus, addressing media not only means encountering and challenging, but also locating them, and, at the same time, situating the analysis of media, which has to confront 'its middling, muddling location in the midst of media' (p. 18). A situational approach is also a relational one (Massumi, 2014): addressable media show that space cannot pre-exist its relations, but that contexts are continuously re-contextualised through them. So, as the spaces of a distributed materiality (the so called Internet of Things) and distributed information (usually termed ubiquitous computing), technospaces do not stand still: they continuously happen.

The attribution of a networked character to every sociotechnical formation is also an efficacious way to escape the rhetoric of substitution as well as that of novelty concerning information society, with all its fears and hopes. The more the planes of our reality interface with one another through locative and ubiquitous media networks, the more difficult it becomes to distinguish between places that are only physical and places that are only virtual. The real/virtual dichotomy becomes inadequate since all experiences are real, but differently mediated. Places are not dead because of the passage of information and communication flows, but simply because 'pure' uninformed places have never existed. Contemporary media, becoming increasingly mobile and locative, rather highlight different ways of performing them, which comprise material (the possibility of transporting mobile devices inside and through physical places) and symbolic elements (the possibility of moving through different forms of even virtual proximity).

In this respect, the recombinant perspective usefully paves the way to a change in paradigm, and the adoption of a performative, non-representational framework when talking about locations inside technospaces. Locations actually take place through conjunctive and disjunctive sociotechnical mediations. At the same time, once the relational quality of locations stretching beyond existing spatial forms is foregrounded, their generativity can also be taken into consideration. As Joshua

Meyrowitz (2005) puts it, 'we do not always make sense of local experience from a purely local perspective. Various media give us external perspectives from which to judge the local. We may be mentally outside, even as we are physically inside' (p. 22). 'Locals are *localized*, places are *placed*', echoes Latour (2005, p. 195). Conversely, to paraphrase Latour (1993) again, even a longer network remains local at all points (p. 118): outside of a local there is always another local, dislocated though it may be.

After re-contextualising the local, thus, it is also necessary to understand how this locality has been generated and redistributes itself along multiple networks. In any case, the consideration of the generativity of locality cannot be measured with dimensional parameter of scale. Scaling is a representational practice among others working to create distinctions, such as between the local and the global, which are presented as foundational when they are really a consequence of observed phenomena. According to a topological understanding, scale is not a fixed variable existing independently from the activity of *scaling* (Latour, 2005, p. 184). As Sallie Marston, John Paul Jones, III and Keith Woodward (2005) affirm, there is no need to resort to extrinsic measuring systems like that of scale to account for sociospatial processes (pp. 424–5) as 'discussion of the site's composition requires a processual thought aimed at the related effects and affects of its *n*-connections' (p. 425): the local and the global do not pre-exist pre-assigned spatial divisions, but can be interpreted as 'the "inside-of" and "outside-of" force relations that continuously enfold the social sites they compose' (p. 426) – which recalls the operations of the baroque interface described in Chapter 1 (see Munster, 2006; Murray, 2008).

Mizuko Ito (1999) defines 'network locality' as the blurring of local 'places' and global 'forces' in the contemporary media environment. This is a geographically extended locality constituted through technological networks, the stretching of which cross-cuts the local/global, location/mobility dichotomies while revealing both the situated character of digital technologies and the networked aspect of geographical places. 'Locality is unbounded and dynamic, an ongoing partial achievement' requiring constant engagement, although, at the same time, always 'grounded in particular social practices, materialized texts, placed infrastructures and architectures'. According to Ito, there is no need to distinguish between network and geographic locality since, in technospaces, 'all localities are ultimately hybrids of geographically and technologically placed connection' (p. 21).

Locality, here, obviously ceases to be just a dimensional term, as its aspects of partiality, relationality and dynamism make clear. This also has important political consequences: the deconstruction of spatial hierarchies frees social actors from the constraints of predetermined positions, usually local ones in terms of contained, more controllable dimensions (Marston, Jones and Woodward, 2005). The openness of technospaces can be thus productively engaged with, being in turn generative of alternative spatial formations (Featherstone, D., Phillips and Waters, 2007). Obviously, not all these formations are necessarily revolutionary, since localising processes can also follow reactionary or colonising routes: namely, 'there is no universal politics of topographic categories' (Laclau and

Mouffe cited in Massey, 2005, p. 165). However, the recombinant perspective foregrounds the role of different agencies in the creation of technospaces, and it contributes to putting an end to such hegemonic narratives which contend that a neat distinction between global and local dimensions, and a conflation of locations and identities, corresponds to spatially bounded forms of intervention (see Featherstone, D. 2007). Indeed, as Massey (2005) argues, responding to Michael Hardt and Antonio Negri's (2001) critique, 'both the romance of bounded place and the romance of free flow hinder serious address to the necessary negotiations of real politics' (p. 175). The contrary happens when location is seen as a productive, intensive dimension, as it is argued by the various declinations of feminist politics of location. In fact, long before the need to reconceive our sense of place and space more fluidly in the current debates about new media, feminist theorists (see Massey, 2005) have shown the dangers of thinking about places and identities as stable and fixed locations.

To give (only) one example, the consideration of place as home, frequently conflated with the feminine, has revealed itself to rest on particular gender and power relations whose boundaries have, in turn, been strictly defined according to such a conception (see Shields, 2006). Massey (1994) notes that the characterisation of place as home usually comes from those who have left, whereas it would be much more interesting to see how often this idea takes shape around those who have been left behind to personify the subjects who do not change, and who are not even allowed the desire of mobility. In her renowned essay on Impressionist women painters and the spaces of femininity in late nineteenth-century Paris, the feminist art historian Griselda Pollock (1988), for example, focusses on the asymmetries of being a man and being a woman in that epoch and how 'the social structuration of sexual difference' would, in turn, influence the way women painted space and were painted in relation to space (pp. 247-8). In particular, she argues that canonical Modernism established itself on a set of gendered practices organised around a number of key 'markers' (specifically leisure, consumption, spectacle and money) on the basis of which different experiences of space could or could not be made. So, for example, if the figure of the *flâneur* embodied the privilege of mobility inside the modern city, a feminist correspondent could not exist in the same terms (see Shields, 2006).[1]

The polarity between the security of fixed locations and the freedom experienced by those who travel, typical of Western Modernity, frequently resurfaces in our contemporary scenario (Kaplan, 2002; Kesserling and Vogl, 2013), albeit on a wider scale. In fact, whereas Modernity was grounded on a metaphysics of presence which produced a 'hypertrophy of the perception of *where*' (Stone, 1995, p. 90) – the subject, as the true site of agency, should always be locatable in order for the law to

---

1 For a slightly different approach, see Bruno's (2002) discussion of the figure of the nineteenth-century travel lecturer Esther Lyons who, although from a privileged socio-economic position, attempted to wear the mask of the explorer in order to expand the horizons of contemporary femininity.

function – the employment and development of new ICTs, the changes introduced by a post-capitalist transnational economy, and the emergence of hybrid physical-virtual mobilities seem to delineate a different scenario. That the self-evidence of the *here* has disappeared, however, does not mean that the *where* has vanished too (Haraway, 1991; Stone, 1995). Rather, mobility and location are increasingly difficult to disentangle: in technospaces, mobility is not equal for everyone, but different mobility regimes emerge (Kesserling and Vogl, 2013) depending on *where* one is, can or wants to be located, and surely some forms of mobility continue to be unaccounted for or cut off from the most glossy global scenario (see Kaplan, 1994).

3.2   Tracing Women's Routes

The revitalisation of the spatial vocabulary used by some feminist thinkers aims to avoid the double bind of the place-boundedness/placelessness dichotomy and to account for a pluralised and hybrid idea of mobility (see Soja and Hopper, 1993). Although there is an intrinsic, and problematic, relationship between travelling and the construction of masculinity, women are also interested in the destabilisation of material and discursive boundaries, as Janet Wolff (1993) has noted. Feminist politics of location, however, invites us to keep in mind that 'destabilizing has to be from a location, and [that] simple metaphors of unrestrained mobility are both risky and inappropriate' (p. 191; see also Haraway, 1992).

Being unbounded does not mean being ungrounded (see Braidotti, 2006), not even today: only a situated cartography can account for such a multiplicity while at the same time embedding critical practice in a situated perspective that avoids generalisations and abstractions. A situated cartography cannot exist as a closed, separate system of representation, nor can it portray a single truth; in fact, remaining open to different social dynamics and possibilities, it works not only as a descriptive, but as a transformative, enactive media (see Haraway, 1991; Braidotti 2002; Massumi, 2014).

The video essays of the Swiss artist and curator Biemann can be seen as feminist cartographies that counter the abstracting tendencies of the rhetoric accompanying mainstream uses of new ICTs with situated accounts that reveal the different roots and different relational networks implicated in transnational mobility. According to Jörg Huber (2003), the practice of video essayism works at putting into relief a set of connections. Firstly, it links theory and practice since it manifests the ways in which theory is embedded in its contexts of production and shows the processes that theories concretely set in motion. Secondly, it appears as both a mobile tool and a means for moving the audience which traverses and translates the world rather than framing it in static pictures. Lastly, and perhaps most importantly, it combines a transdisciplinary quality with a self-reflexive stance in that it unmasks the position of the speaking/viewing subject while also accounting for its relational character. All of these features render the video essay an appropriate instrument for tracing the intersections of location and mobility in the transnational scenario.

In her video essays, Biemann shows the interplay of the material and symbolic effects produced by the flows of transnational economies and ICTs. Although ICTs open up routes and alternatives previously unimagined, they also tend to be used as instruments of control for the reinforcement of existing physical and virtual borders. Biemann intends to account for the multiple locations of transnational actors in order to disclose the complex interrelations between asymmetrical mobilities across several borders and, at the same time, reveal the embeddedness of the technologies they both talk about and utilise.

Whereas in Second Modernity[2] mobility has potentially become a 'horizontal' condition for everyone, affecting both social practices and imaginaries (Kesserling and Vogl, 2013), location and mobility do not have the same meaning for everyone. The 'post-national' condition of the refugee, for example, which Biemann describes in her video essay *X-Mission* (2008), exerts a profound change both on the refugees' self-perceptions and how they are perceived as human beings, a perception that is filtered through the politics of the Nation State and the visual rhetoric of humanitarianism conveyed through media. So, transnational subjects like refugees or sex workers do not experience the kind of mobility theorised by those who believe that, in our techno-driven world, the act of travelling no longer requires any kind of material displacement. Avoiding the rhetoric of similar narratives and asking 'who suffers, who troubles, who works these technologies of travel' becomes paramount (Kaplan, 2002, p. 40).

As Biemann's artworks make clear, these questions also imply the need to reconceive the act of visualising and representing (women), and thus to rethink the function of artistic practice and the role of the artist as witness/author. In order to become an 'embedded artist', as Biemann claims to be in her work *The Black Sea Files* (2005), does it suffice to collect information and comment on what one sees? Or is it not imperative to adopt a self-reflexive stance so that the location of the video essayist itself emerges through the proposed meanings? 'Knowledge from the point of view of the unmarked', says Haraway (1991, p. 22) is pure fantasy, built on the rational myth of *everywhereness* which, from a feminist situated perspective, turns out to be the same as *nowhereness*.

Discussing curatorial practices in postcolonial sites, Biemann (1997; 2003) notes that already posing such questions is a good entry point for transforming existing power relations without simply reproducing them. In fact, Biemann's 'navigational' rather than representational video essays (Biemann, 2003) do not mirror existing spaces from above, portraying the world in a fixed structure of

---

2   With the term Second Modernity, Kesserling and Vogl (2013) refer to the changes in mobility introduced in postindustrial societies. Whereas in First Modernity mobility was important but still restricted to certain social actors, remaining basically a material issue, the link between communication and mobility, and the increasing feasibility and availability of ICTs in the late twentieth century has transformed mobility into an everyday phenomenon which, although differently, now involves multiple actors, deeply affecting social norms and traditional categories of analysis.

power and meaning. Instead, they locate 'the space of theorizing' (hooks cited in Kaplan, 1994, p. 143), as well as that of visualising, within the complex system of signification that images can only partially render, unmasking the function of the actual instruments that are employed in their visualisation. This is how the artist describes her choice of the video essay as her privileged artistic medium:

> like transnationalism, the video essay practices dislocation; it moves across national boundaries and continents, and ties together disparate places with a distinctive logic ... The narration in my video essays – the authorial voice – is clearly situated, in that it acknowledges a very personal view. This distinguishes it from a documentary or scientific voice. Though the narration is situated in terms of identification (as it is articulated by a white female cultural producer), it isn't located in a geographic sense. It's the translocal voice of a mobile, travelling subject that does not belong to the place it describes but knows enough about it to unravel its layers of meaning. The simple accumulation of information and facts for its own sake is of little interest to this project. My video essays are not committed to a belief in the representability of truth. Rather, my intention is to engage in a reflection about the world and the social order. This is accomplished by arranging the material into a particular field of connections. In other words, the video essay is concerned, not with documenting realities, but with organizing complexities. (2007, p. 130)

In this respect, Biemann uses video essays as dislocating media for *diffracting* rather than simply representing what they visualise (see Haraway, 1997; Barad, 2007). Not looking for a final truth behind images, they rather follow the connections that emerge through the adoption of a partial perspective, elaborating upon the specificities, the complicities and the differences of our ways of seeing (Haraway, 1991, p. 190). In other words, Biemann's cartographies do not simply trace but also perform the connections between embodiment and movement, location and displacement, delineating a situated opportunity to articulate our contemporary imaginary, particularly when a discussion of new technologies is involved.

It is no coincidence that most of the contemporary figurations of hybrid subjectivity discussed by Braidotti (1998), such as the mail-order bride, the rape victim of war, the au pair girl and the *doméstica*, along with the techno-skilled cyberfeminist, are the very protagonists of Biemann's video essays. As Rob Shields (2006) puts it, discussing the re-spatialisation of cyborg bodies in the contemporary scenario, 'sites such as Home, Market, Paid Work Place, State, School, Clinic-Hospital and Church deserve to be re-mapped for their politics at a nano- and biotechnical scale without assuming that they are inhabited by a unitary, integrated and self-coherent political subject' (pp. 217–18). At the same time, we need to ask how these new forms of subjectivity are displaced and redistributed as well as constituted along intertwined material and digital networks.

In the trilogy comprising the video essays *Performing the Border* (1999), *Writing Desire* (2000) and *Remote Sensing* (2001), Biemann focuses on what links women's mobility to new technologies, following the routes of globalised and feminised labour. As Saskia Sassen (1998) has noted, the internationalisation of manufacturing, which creates a sharp polarisation between the superprofit-making capacity of corporations and the feminisation of the kinds of jobs upon which high-income gentrification draws, while rendering women the invisible subjects of this economy, also alter previous gender hierarchies and give women a new kind of control over their self-awareness, mobility and money, leading to new forms of female solidarity and transnational alliances.

To shoot *Remote Sensing* (2001), which focusses on the organised and individual paths of and reasons for the global sex trade, Biemann travelled to some of the places where the global sex industry has flourished, such as the Thailand-Myanmar-Laos triangle, the border between the former German Democratic Republic (East Germany) and what was once Czechoslovakia (now the Czech Republic and Slovakia), the United States-Mexico border, and the former US Marine bases in the Philippines. Ultimately, however, *Remote Sensing* also explores some locations that are entirely imaginary.

Figure 3.1   Still from *Remote Sensing*, video essay by Ursula Biemann, 2001

Source: Courtesy of Ursula Biemann.

On average, almost half of the total population migrating every year is female (United Nations, 2013). Women migrate for many different reasons, some of them leading to, or connected with, trafficking: the entertainment industry, sex tourism, forced prostitution, discrimination, political instability, the need for complementary incomes, the supply of family services where these are lacking, or the social restructuring of gender relations. Women may also desire to move to affluent countries to realise their personal dreams.

In *Remote Sensing*, Biemann makes wide use of satellite images, inviting us to *feel* rather than merely see them. Vision, recontextualised as a situated practice, dismantles the fiction of the neutral distance of technoscientific methods (Kwan, 2002; 2007; Parks, 2005). The same satellite instruments used to keep women's movements under control also produce new perspectives that render the visual field more complex. If, on the one hand, some kinds of visualisation devices interfere with women's mobility, limiting their actual movements by tracking their routes across borders, technologies such as the Web on the other hand help women to become transnational actors – as occurs in the case of the cyberbrides – giving them a greater opportunity to visualise and fulfil their desire for mobility.

As the narrative of *Remote Sensing* explains, viewing sex migrants exclusively as victims reinforces sexist stereotypes, leading to the creation of restrictive measures intended to prevent mobility. Thus, a careful consideration of the kinds of power-related differences at stake in these movements, as well as their possible intersections, is imperative (Verstraete, 2007). Several national economies have become increasingly dependent on the remittances of sexualised mobilities, showing the interdependence of location and dislocation in the global scenario. Women who move along sexualised routes, however, frequently create their own alternative economies and 'circuits of survival'.

In *Remote Sensing*, electronic travel schedules indicating women's journeys as well as their personal data – timetables, departure and destination places, longitude and latitude coordinates appearing alongside information regarding age, height, weight, ID and visa numbers – scroll over the video images. In some instances this information appears at the side of the screen, while in other it is superimposed upon the x-rayed portraits of the travelling women, against a backdrop of unrecognisable landscapes. Here, Biemann mimics the official devices used to classify and depersonalise their identity, fixing them in a taxonomic grid of detached observation. In addition, she makes frequent use of the split screen, juxtaposing apparently frozen satellite images with images of movement recorded 'from below'. The narratives of the interviewed women – NGO activists as well as ordinary women, whom we sometimes see and sometimes only hear off-screen – *embody* these otherwise mute and anonymous images.

Although the video contains a wealth of information and seems to adopt documentary tools, its intent is not properly speaking representational: rather, Biemann investigates the interplay between the symbolisation of the feminine and the groundedness of women's lived experience. Here, the private and domestic sphere, where women have traditionally been confined, and the economic and public sphere, usually considered a masculine domain, appear to overlap, re-signifying and expanding the space of the feminine itself. The women interviewed tell stories that reveal different backgrounds, intentions, interests and desires, thus disrupting the flattening logic of the GIS (Geographical Information Systems). By looking at the negotiations occurring among the abstract flows of information, money, and representations, and the material flows of people, we as spectators confront the issue of women's mobility in a subtle way, avoiding the binary opposition of passive victim vs. free agent (Biemann, 2005, p. 185).

A very specific example of how women's mobility can benefit from the employment of new technologies is offered by *Writing Desire* (2000), a video essay on the dynamics of desire and the new female subjectivities generated by the different uses and locations of ICTs. Here, Biemann focusses on the phenomenon of the mail-order bride market on the Internet, particularly common in postsocialist and Southeast Asian countries. *Writing Desire* shows the exchanges between virtual and physical bodies in cyberspace and suggests the way in which virtual instruments facilitate women's mobility by linking virtual and real migrancies. In fact, the women's imagined – but not imaginary – places (van Alphen, 2002) always maintain a connection with their real locations, which in turn come to be experienced differently according to the diverse symbolic projections.

Writing Desire opens with a glossy image of a tropical beach while statements about the passivity of women 'waiting to be rescued' by foreign capital flow across the screen. A parallel between women and nature is established, with all the associations that this implies: above all, the idea that women, like nature, are essentially immobile and passive, and outside history (Williamson, 1986) as opposed to men, who can move, and thus change, continuously. 'The body signifies the anonymous exotic, the desire to be conquered', says Biemann. But this nostalgic image only apparently contrasts with the electronic fantasies that are articulated immediately afterwards. In both cases, what stimulates fantasies of virtual bodies are highly coded icons, whose interplay of distance and proximity generates 'a sense of always approaching, but never reaching'. In this instance Biemann does not merely show how women are signified as desirable bodies; she also shows us how women signify their desire.

In *Writing Desire*, diverse writing positions, in fact, coexist without creating a dichotomy between female subjectivities in purportedly 'advanced' Western societies, where women follow a postmodern logic of desire and adopt a free and easy approach to sexuality, and those 'third world women' who, in their fight to survive, are obliged to offer their care and services in order to escape from degrading living conditions. Maris Bustamante, for instance, not only writes her desire, but also realises it. This 50-year-old Mexican woman, feminist, artist, widow and university professor looks for a partner online because she is not satisfied with the men that are available to her in the local Mexican environment. Finally, she meets John, a lieutenant colonel in the US Marine Corps, and marries him. Although this choice may seem unexpected, they manage to form a new family, embodying their virtual fantasy in daily experience.

Studying a group of well-educated middle-class Mexican women from Guadalajara seeking a transnational marriage in the USA, Felicity Schaeffer-Grabiel (2004) argues that these women look for men 'over there' in order to leave behind traditional restrictive values associated with womanhood and to improve their lives. When interviewed, most of the women state their intention to escape from the 'machismo' of Mexican men, simultaneously revealing their critique of the national body, which devalues women's changing roles in private and public spaces. Even so, they tend to project a self-image mirroring the ideal of a traditional Mexican woman who is suitable for an American man looking for an authentic 'precapitalistic' marriage. This highlights, once again, an internalisation of stereotypes and a simultaneous re-enactment and transgression of gender roles. Like Maris Bustamante, these women have attained a consumer position, although their achievement is often ambiguously tied to the commodification of their own bodies.

If *Writing Desire* shows how women can overcome phantasmatic and real borders through new ICTs such as the Internet, *Performing the Border* (1999) deals with the reciprocity between the gendering and the technologisation of female subjectivities as they negotiate the contradictory dynamics of transnational space. ICTs and visualisation technologies structure the technologies of gender (De Lauretis, 1987; Terry and Calvert, 1997; Volkart, 2000), but they too are gendered as well as racialised, caught in a complex network of socio-historical relations. In order to understand how and why the gendering of a technology does not necessarily occur to the detriment of women, we should also consider who is empowered in the deployment of a specific technology, where and for what reason, beyond the mere question of access. In seeking to answer these questions, attention must be paid to the 'implicit and explicit socio-cultural hierarchies within transnational urban work spaces shaped by the ICT related technology work' (Gajjala and Mamidipudi, 2002).

In *Performing the Border* the geographical border becomes a powerful figuration for an analysis of the performativity of several boundaries: those between masculinity and femininity, the organic and the machinic, production and reproduction, location and mobility, the real and the virtual. Of course, borders do exist, but they are neither natural nor fixed. They are differently and constantly re-signified by people crossing them, either sanctioning or transgressing their logic (see Zanger, 2005).

Another of Biemann's video essays, *Europlex* (2003), made in collaboration with the anthropologist Angela Sanders, looks at the multiple movements generated by transnational economies along the border, this time between Spain and Morocco, and in the Spanish enclaves of Ceuta and Melilla (North Africa) in particular. The video narration is divided into three *Border Logs*. *Border Log I* follows women smuggling goods under their dresses to Africa, a project that requires them to move illicitly across a geographical border. *Border Log II* describes the daily routine of African women going to work as *domésticas* in the enclaves, having to commute between time zones, thus crossing a more 'invisible' border. Finally, in *Border Log III*, we see Moroccan women working in sweatshops inside the transnational area. Their experience of the border is even subtler here since they only commute between different cultural environments; nonetheless, they experience a continuous shift between these environments and, in so doing, they perform yet another kind of border.

The narrative focus of *Performing the Border* is the experience of the young women working on assembly lines in plants situated in the export processing area of Ciudad Juarez, between Mexico and the USA. Here, the North American Free Trade Agreement (NAFTA), a trilateral agreement signed by Canada, the USA and Mexico in 1994, along with its subsequent reinforcements, is one of the 'governmental apparatuses' articulating the contradictions of the nation state with the development of global capitalism (Mezzadra and Neilson, 2013, p. 53), which make the border a very contradictory zone and its crossing more and more regulated.

Paradoxically, the microelectronic components assembled in the Juarez *maquiladoras* are used to produce technologies for information processing, satellite systems, and optical instruments, which are often the same technologies that reinforce existing borders and maintain control of women's bodies, literally and metaphorically. In fact, visualisation technologies track women's movement in space as if they were commodities, but they also circumscribe their gender identity according to the transnational standards of production and consumption. Women in Ciudad Juarez live a boundary condition since they perform the border and embody all the anxieties it evokes; these are related to national and colonial fantasies of mastery and domestication in which their geobodies signify the traditional values of the motherland as well as the transnational logic of the corporate economy. In both cases, the abstraction of women's bodies from the actuality of their lives renders them vulnerable subjects, suspended between the coalescing forces of the natural and the technological, which cooperate to keep them under control.

As Berta Jottar, drawing on Gloria Anzaldúa (1987), states in the opening of the video, the border is a wound, a 'surgical place' requiring constant healing. The possibilities of the border, both a corporeal and territorial confinement, are variously performed: these range from squatting inside houses built on the remnants of industrial wastes to, in some cases, trying to run away, possibly with the help of *coyotes* like Concha, who helps pregnant women steal across the border in safety so that they can give birth in a US hospital. The sex industry, by now a structural component of the global economy, also flourishes here, often because of the women's need to produce additional income. This has gradually led to the emergence of an entertainment industry that addresses women as autonomous consumers of leisure activity, affecting their relationships and impacting on their role in society. At the same time, however, sex work remains the only possibility that young women living in this area have to make ends meet if they are not educated enough to enter a *maquiladora* or lack the references to work as *domésticas*.

In this as in the other videos, Biemann uses what Volkart (1999) has called a 'flow discourse', a fluid aesthetic technique with the camera constantly moving between subjects and places. The sequences are frequently shot from a moving car, and some of them are slowed down, out of focus shots. This fluidity, however, is never linear. Rather, it breaks into multiple perspectives, which are shown next to each other (in split screen), inside each other (the use of multiple windows), over or under each other (the shift between perspectives from above and from below). These displacements include the video essayist's position too, when Biemann alternatively speaks off-screen or writes her 'working' notes over the image. All these techniques, achieved for the most part during the editing stage, create sutures that subtract the video essays from the logic of both authorial and spectatorial comprehensive vision, preventing a unified perspective, as Alexandra Karentzos (2013) also notes about the montage techniques in Lisl Ponger's film work, an artist who analogously works on the ambivalences of mobility and the complicities of representational practices. In fact, Biemann's video essays are figurations of the 'grey areas in-between' that the activist Bandana Pattanaik refers to in *Remote Sensing*: it is in these zones that the video essay captures the otherwise invisible flows of the 'geographies of survival' (Biemann, 2007) and returns them to visibility.

Biemann's cartographies contemplate nuanced, shifting perspectives that do not rely on the transcendent logic of binary optics. The *ecology of visuality* to which the artist appeals (2008) parallels the call of feminism for an ethics of geospatial practices (Kwan, 2007; see also Schuurman and Pratt, 2002; Sui, 2004; Propen, 2006; Pavlovskaya and St. Martin, 2007; Le-Phat Ho, 2008). She believes in a 'sustainable representation' that does not simply reproduce or reflect a pre-existing reality but reveals itself as an instrument of interpretation and navigation, disclosing the various ways in which spatial discourse 'takes place' together with the acts of observing its taking place (Huber, 2008, p. 173). This entails taking into account alternative uses of both space and (its) representations, where various forms of agency, including critical agency, are seen as contributing to the geopolitics of social formations and their discursive practices. Such visual ecology contends that observing is not a neutral act that only registers what the eyes see. Observing is performative, and it requires that the observer be aware of the power dynamics and the complicities of vision, how differences are made and not given, and of the intra-actions between the observer and the observed (Barad, 2007).

Barad (2014), relating her own and Haraway's theorisation of diffraction to both Anzaldúa's theory of the 'mestiza' and Min-ha/Haraway's notion of the 'inappropriate/d other', notes that what links them is the idea that 'boundaries don't hold; times, places, beings bleed through one another' (Barad, 2014, p. 179). The wound that crosses the US/Mexican Border, then, also relates the subject and the object of knowledge, cut together *and* apart in the same activity of connection and separation. A diffractive methodology articulates and goes through such cuts rather than suturing and concealing them.

Trinh T. Min-ha (Gržinić and Min-ha, 1998) observes that, nowadays, technologies are too often used to access a reality without mediation, in order to pursue an aesthetics of objectivity. When this occurs, reality becomes that which is immediately visible, while media simply disappear. However, the removal of boundaries, whether by rendering them invisible or blurring them, is, in the end, illusory. Rather, Min-ha says, 'it is a question of shifting [the boundaries] as soon as they tend to become ending lines'. Here again, the issue of the inappropriate/d other comes into play: 'to be inappropriate/d', echoes Haraway, 'is not to fit in the taxon, to be dislocated from the available maps specifying kinds of actors and kinds of narratives, not to be originally fixed by difference' (1992, p. 299). Unfixing differences is, thus, an act of responsibility and commitment, as it means abandoning the unbridgeable distance of absolute difference and recognising the involvement of the subject in the field of the object, a field whose self-sufficient totality must be questioned starting from the idea of visual distance that it relies upon.

3.3   Navigational Maps

The emergence of a shared sensibility in the arts and sciences and the construction of the world as a visual totality at the end of the nineteenth century are at the core of Gregory's *Geographical Imaginations* (1994). He addresses the problematic of visualisation in relation to the constitution of modern geography in the first chapter, which is significantly entitled 'Geography and the World-as-Exhibition', after a renowned essay by Timothy Mitchell (Mitchell, T. 1989). Gregory draws on elements of the history of art, science, and other disciplines to analyse the ocularcentrism of modern geography, but his arguments can easily be reversed and used to understand the role of visualisation technologies and such 'geographical' tools as cartography and mapping in general in the arts and sciences.

Gregory (1994) introduces Mitchell's essay in the second part of the chapter, noting how Mitchell talks about the ordering of the world as exhibition in Western Europe during the nineteenth century. Mitchell draws on an episode in which a delegation of Egyptian scholars invited to the 1889 International Congress of Orientalists in Stockholm faced the construction of alterity as exhibition twice: firstly, when they visited the Egyptian Pavilion at the World Exhibition in Paris during their voyage; secondly, when they themselves were treated as Orientals, rather than Orientalists, at the congress in Stockholm. This *mise-en-scène* corresponded to, and at the same time relied upon, a process of objectification of the things displayed that was made possible by a distant viewing subject constructed as spectator throughout this viewing modality. As in the case of panorama paintings or dioramas, this caused a paradoxical belief in the realism of representation (that is, its correspondence to an external reality), and a simultaneous acknowledgement of the exhibition *as* representation, 'set up for an observer in its midst' who was simultaneously surrounded by images yet excluded from the order of display

(Mitchell, T. 1989, p. 223) – the point of view being created precisely through such a distancing move[3] (see also Bolt, 2004, p. 25 ff.).

In T. Mitchell's (1989) analysis, this initiated a labyrinthine play of cross-references and mirror reflections between reality and representations, and the effect of an (endlessly unreachable) 'external reality' that extended well beyond the proper space of the exhibition to include the entire city of Paris, exemplified by the shopping arcades, perceived by Eastern travellers as miniaturised worlds. However, if people travelling from Asia to Europe perceived the European world as an exhibition, clearly organised and composed for the viewer to see, Europeans travelling abroad suffered from the 'absence of pictorial order' (p. 227) of foreign places. This was, for example, the case in Flaubert's (1983) account of Cairo, where no visual distance (and no visual 'hygiene') was allowed for the stranger; thus, Mitchell notes, no comprehensive vision from any 'position set apart and outside' (that is, the constitution of the point of view) was possible (p. 229). 'Strangeness' was 'expressed in terms of the problem of forming a picture' (p. 228), even though several stratagems were eventually found by European travellers in order to artistically recompose the experience of their journey.

What T. Mitchell (1989) highlights is how the organisation of the world as exhibition created the belief in two different counterposed realms, the real world and the represented one; reality, in consequence, turned out to be what could actually be properly represented, that is, recomposed as an exhibition, as a distinct object, for the beholding eye of the viewing subject (see also Bolt, 2004). As he continues, 'what matters about this labyrinth is not that we never reach the real, never find the exit, but that such a notion of the real, such a system of truth, continues to convince us' (p. 236). Drawing on T. Mitchell's analysis, Gregory (1994) traces the genealogy of the world as exhibition back to the role of linear perspective and down to the invention of the *camera obscura* and the colonising moves of the nineteenth century. It is no coincidence, for Gregory, that both the World Exhibition in Paris and the International Congress of Geographical Sciences opened in 1889, nor that many types of optical machinery were invented in that same period, paralleling the institutionalisation of human geography (pp. 38–9).

Some of the most commonly adopted techniques of spatial construction, such as *perspectiva artificialis*, but also certain genres, such as landscape painting and *views*, and specific aesthetic categories, such as the picturesque, surely acquire a different significance when they are related to the appropriation of space going on from the Renaissance to the colonial enterprises of the nineteenth century (Duncan, C. 1982; Williamson, 1986; Solomon-Godeau, 1989; Nochlin, 1992; Rose, 1993; Gregory, 1994; Pacteau, 1994; Nast, 1998; Bruno, 2002). Let us consider, for example, the continuous exchanges between landscape painting and cartography: and it is no coincidence that both the translation of Ptolemy's *Geography* from the

---

3   Mathematically speaking, mapping is precisely the correspondence between two sets in which each element of the first one has a counterpart in the second one (see Farinelli, 2003, p. 78).

Greek into Latin and Brunelleschi's first experiments with linear perspective took place in Florence at the turn of the fifteenth century, testimony to the new desire to know and master space that was beginning to emerge (see Rees, 1980).

However, that a comprehensive visual field existed means that a series of excluded visualities also existed at the limits of such a totality (see Deutsche, 1995; Bruno, 2002). In fact, as Gregory (1994) recognises, other scopic regimes in Western Modernity – and clearly other Modernities, each with their different visualities – challenged the idea of the world as a single visual totality, although their power was less visible or, we could also say, their visibility was less powerful than the prevailing idea of the world as a visual whole. Even so, the representational imaginary emerged as the dominant paradigm of Western Modernity (see Foucault, 1994) because it was the most suitable to support the constitution of the identity of the Westerner as an autonomous and self-reliant individual, endowed with the (visually dependent) ability to calculate, categorise, separate and identify the Other for its own purposes.

That is why T. Mitchell (1989) affirms that Orientalism (see Said, 1978), before being an aspect of colonial domination, which it surely was, was first of all 'part of a method of order and truth essential to the peculiar nature of the modern world' (p. 236). One only need think of the role of photography in the visual construction of racial and gender differences supporting the project of the nation and its colonial expansion that so clearly emerges, for instance, in the rich survey of pictures put together by Coco Fusco and Brian Wallis (2003) on the occasion of the exhibition *Only Skin Deep. Changing Visions of the American Self.*

Besides artists and art historians (see Berger, 1972; Pollock, 1988), for obvious reasons, the analysis of the unevenness of spatial imaginations has mostly come from the scholarship of feminist geographers (see Gregory, 1994; Deutsche, 1995; Robinson, 2000). On the one hand, feminist geography has widely relied on the feminist critiques of representation and visuality developed in visual and film studies (Rose, 1993; 1995a; 1995b; Gregory, 1994). On the other hand, visual studies have also largely employed geographical and spatial concepts for their analyses of the relations between the technologies of vision and the imaginary of mobility and location (Friedberg, 1993; Rogoff, 2000; Bruno, 2002; Gordon, 2010). In this context, for example, one of the geographical concepts *par excellence*, that of field, has undergone a profound critique, something that can reasonably be compared to the reflexive turn that, in the 'hard sciences', has dismantled the traditional divide between the observer and the observed and instead prompted attention to the observer's position (Latour, 1987; Haraway, 1991; Rose, 1993a; 1993b; Nast, 1994; Staeheli and Lawson, 1994; Sparke, 1996; Harding, 2004a).

In fact, the notion of field, which is shared by different disciplines, from geography to anthropology to the military and the visual arts, implies the existence of a place to be penetrated, explored and possessed, which, additionally, is frequently characterised by gendered metaphors. The naturalisation of the field not only presupposes that the field is a passive, a-historical space 'out there', but it also makes the space of the observer transparent, which is to say 'innocent', thus eliminating any intra-action (Barad, 2007) between differential positions inside the same field, as well as the co-constitution of the subject and the object of observation. Problematising the field, then, means blurring the distinction between 'the politics of fieldwork' and 'the politics of representation' (Nast, 1994, p. 57), and acknowledging the performativity of visual practices.

After retracing the genealogy of dominant visuality in Western Modernity, Gregory (1994) extends his analysis to the problematic of representation in relation to contemporary time-space compression. As he shows, it is around the middle of the twentieth century, when geography is finally constituted as a formal spatial science, that the idea of the world as exhibition reaches its apotheosis. The alternation of 'framing' and 'deframing' as the forces of composition and transformation that Elizabeth Grosz (2008) considers the condition of all the arts, appear to incline towards a prevalent territorialisation dynamic. The fascination with the grid, which Rosalind Krauss (1985) identifies in Modern art as well, permeates geographers' interest in spatial structures. A metageographical model based on the grid, or what Gordon (2007) calls 'the graticule',[4] still prevails in early narratives of cyberspace. As knowledge and space become more and more associated, their colonising and disciplinary power is reciprocally reinforced and increased (Gregory, 1994, p. 63).

Clearly today, the role of new information and visualisation technologies in the perception and experience of space has changed our relation with the world, which seems now to be *travelled through*, rather than simply *gazed at*. Discussing contemporary cinema, for example, Giuliana Bruno (2002) distinguishes between the visual experience of *sight-seeing* from the more recent experience of *site-seeing*, implying physical involvement, immersivity and polysensoriality (see also Friedberg, 1993; Gemini, 2008). However, we should be wary of attributing this feature to new technologies and wonder whether purely visual media have ever existed or if it is not that the visual has undergone a process of 'purification' from the contamination with the other senses (Mitchell, W.J.T. 2005, p. 260). This is all the more true as, firstly, media cannot be considered in their isolated 'specificity', as purely material technologies, but are always in conjunction with situated, hybrid practices. This is why visual culture does not take vision for granted but questions the very 'nature of visual *nature* – the sciences of optics, the intricacies of visual technology, the hardware and software of seeing' (Mitchell, W.J.T. 2005, p. 264). Secondly, media have never existed as such, being separated by the events of mediation of which they are temporary 'fixings' (Kember and Zylinska, 2012, p. 21). In this respect, it could be said that media work at *dis*-integrating, rather

---

4 'Graticule derives from the Medieval Latin word craticula, which means "little grating". The definition of the word grating is a material used for containment or preventing access. The abstraction of the graticule to symbolize earth, then, might be understood as the abstraction of containment. Very often the graticule has little connection to the actual map, but it is almost always included as a shorthand means of communicating the stable globe as a reference point, and thus a mastery of whatever is plotted therein' (p. 76).

than at integrating society, since they interface with a social that does not exist 'out there', and that media do not represent, but instead enact (p. 31).

Harking back to Gregory (1994), he asks, do we assist in the dissolution or, rather, the apotheosis of the world as exhibition today? In fact, he notes, the three characteristics that T. Mitchell attributes to the world as exhibition still continue to exist. They are: certainty, which pertains to belief in the possibility of representing the real; paradoxicality, which makes us use representations to depict reality, although we feel that we can only access reality through representations; finally, the colonising effect, or the complicity of visual representations in various systems of power (p. 64). For instance, digital media seem to further radicalise the representational dynamic of involvement and distance. In Gregory's (1994) words:

> That sense of being 'within' is greatly enhanced by the visualizing practices of more advanced geographical information systems and interactive telecommunications systems, which at once set the world at a distance – accessed from a platform, seen through a window, displayed on a screen – and yet also promise to place the spectator in motion inside the spectacle. (p. 66)

For instance Benjamin Bratton (2013), in an essay which resonates with Jameson's (1991) mourning for the loss of critical distance, analyses in dystopic terms the 'collapse of representational distance' in the contemporary 'expansion of interfaciality'. Whereas he recognises that the unfolding of interfacial objects into the environment and their increasing interoperability can have the positive consequence of getting rid of a too-narrow anthropocentric perspective, he sees many 'semiotic' technologies, such as augmented reality (AR) applications, as causing the collapse of critical interpretation together with distance, in consequence of a 'self-dispossession of commitment' that would derive from the transformation of the world in a gigantic subtitled screen already fully explicated in front of our eyes.

Indeed, today, the 'video-digital gaze' (Boccia Artieri, 2004, pp. 136–7) does not need to stand apart in an invisible, detached place to work, as it did according to a panoptical model (Foucault, 1977), since observation and self-observation have now become the environment that we move through, and a condition of diffuse visuality and inter-visuality characterises the contemporary moment. Then, the paradoxical aspect, here, does not only lie in the always new combinations of reality and representation, that labyrinthine chain of mirror reflections already proper to modern Western cities and now enhanced by the pervasivity of interfaces in smart environments. It also pertains to the necessary recombination of the visual in a whole series of mediations that, on the one hand, urges us to relationally reconsider both visuality and representation and, on the other, requires a situated and performative approach to both that can account for the various sociotechnical configurations that vision generates. This is why, for instance, Ash Amin and Nigel Thrift (2002, p. 128) as well as Latour (2005, p. 181) note that the other side of the coin of pervasive visibility is of necessity oligoptism, since ubiquitous as they may be, today's networks do not only produce different spatio-temporalities, and they cannot produce all the spatio-temporalities that still remain to invent.

An example of the coexistence of different optics inside networks of control is the phenomenon of *sousveillance* (see also Kitchin and Dodge, 2011, p. 230), a term that indicates a form of inverse surveillance consciously enacted from below, which makes use of portable and wearable wired/networked devices. An example of sousveillance practices is *surveillance art*, which includes actions and artworks of groups like the Surveillance Camera Players, the Institute for Applied Autonomy and the Bureau of Inverse Technology, or yet the diverse forms of *lifelogging*, either continuous or intermittent, allowed by the employment of contemporary mobile devices. The blogger James Bridle has, for instance, created a sort of visual Wikileaks on his website *Dronestagram* (2012–): posting images of places that are the target of drone attacks in Pakistan, Yemen and Somalia on Instagram, Tumblr and Twitter, after confronting, or better, diffracting (Haraway, 1997; Barad, 2007), the Google Maps images with the data of the Bureau of Investigative Journalism, Bridle presents us with the operations in which drones have been secretly employed by the US army causing civilian casualties, whose number is still kept secret. The places that Bridle shows us are very often border zones, or visually insignificant ones, that for these reasons we would rarely get to know. These become valuable for our gaze the moment in which we are put in the same optics of the drones. Ours, however, becomes a very different visual position, one informed as it is by a different awareness of the events we look to.

In accordance with Haraway's (1991) technoscientific critique of totalising and disembodied vision, Gregory (1994, p. 67) believes that contemporary systems of visuality can be interrogated differently, which is to say *diffracted* (Haraway, 1997; Barad, 2007), so that different 'configurations of technopower' become observable in this visually enhanced world. Even if visual technologies such as those used in GIS seem to confirm the heightened role of the visual today, because they are primarily treated as scientific and 'detached' tools of observation, we must be able to discern in this approach a 'rhetoric of concealment' which disguises the fact that these technologies have also been produced and consumed *somewhere* (Gregory, 1994, p. 65; see Haraway, 1991) and can be in turn generative of different experiences.

Gregory (1994) notes how a representational bias still permeates both the idea of representation as simulation and of representation as mirror image; in each case, the interrogation of the performativity of representational practices is foreclosed. It is worth quoting the entire passage here:

> In the center is spatial science, which, in its most classical form, retains the distinction between reality and representation but which, through its consistent focus on *modeling*, does at least draw attention to the process of representation as a set of intrinsically creative, constructive practices. Hence, in part, the accent on the aesthetics of modeling, on the 'elegance' of spatial models (though the metaphoric of power is never far away). Tracking forward and to one side of spatial science are explorations of cyberspace and hyperspace, which often

abandon the distinction between reality and representation as the unwanted metaphysical baggage of modernity and chart instead a postmodern world of representations and simulations. But in doing so they too draw attention to the constructive function of representation ... On the other side of spatial science, however, are spatial advances in GIS which seem to move in precisely the opposite direction: to assume that it is technically possible to hold up a mirror to the world and have direct and unproblematic access to 'reality' through a new spatial optics. The question of representation, of regimes of truth and configurations of power, knowledge and spatiality, is simply never allowed to *become* a question. (p. 68)

Maps are, *par excellence*, media of knowledge, communication and representation of the world (Kitchin, Perkins and Dodge, 2009) and a privileged terrain to examine the differences between a representational and a performative approach to space and media. As Gregory (1994) notes, the 'cartographic anxiety' that derives from not being able to distinguish representation and reality in the relation of the map and the territory does not imply that we either yield to the simulacrum or to dystopian scenarios of universal visuality (p. 67). Asking which comes first, the map or the territory, reveals its uselessness given that both the question of whether a map is an accurate depiction of what is outside, and whether this outside really exists depend on a representational imaginary (November, Camacho-Hübner and Latour, 2010, p. 589). Things appear more complicated than in Jean Baudrillard's (1994) view.

Very often, maps have been 'sealed', so to speak, by a series of codes contributing to the idea that they were not even produced by human hands, their truthfulness only sanctioned by a series of institutions disguised behind impersonal acronyms (Wood, 2006, p. 9). The more maps were precise and accurate, the less their mediation in relation with the represented was visible. Representational cartography used the map in the same way that the Shannon-Weaver theory (Shannon and Weaver, 1963) intended communication, that is, according to an idea of the reduction of error and distortion necessary for the optimisation of data transmission: an exact and instantaneous vision concealed the spatio-temporal intervals (see Crandall, 2006), that is, the chain of mediations, through which maps were produced and employed.

More or less after the mid-1980s, thanks to changes introduced in computer graphics and the emergence of digital geovisualisation as well as critical cartography, the concept of maps has changed from one of simple products to being progressively considered as *situations* requiring an always more interactive participation and forms of shared construction. Critical cartography shifts attention away from the map as object and toward mapping as a set of inventive and enactive practices in which spaces and its representations are co-constituted (see Crampton and Krygier, 2006; Kitchin, Perkins and Dodge, 2009; Massumi, 2014). The ontogenetic quality of maps has been further foregrounded where the 'contingencies and relationalities' of mapping practices (Kitchin, Perkins and Dodge, 2009, p. 15) have been given more attention than maps as products. The easier traceability of the associations of the material infrastructure that produces technospaces (Latour, 2005), as well as some specific features of today's information and communication media, such as localisation and mobility, and the related affordances of navigability and shareability, increase the mobilisation of 'maps'.

There is no need to suspend our lived space and read the represented space on a map today-since the map and the territory are tangibly integrated in an uninterrupted experiential field. Here, the map, rather than being used as a 'wayfinding' or visualisation tool, reveals itself as an interface through which technospaces are unceasingly performed (Gordon and de Souza e Silva, 2011, p. 19 ff.; see also Lemos, 2009) between knowledges, experiences and imaginations.

It must be noted, however, that this interfacial aspect of maps is not new, as it does not directly depend on digital technologies, but is only put forth by them. Actually, as Latour (1987) writes,

> when we use a map, we rarely compare what is written on the map with the landscape ... we most often *compare* the readings on the map with the road *signs* written in the *same* language. The outside world is fit for an application of the map only when all its relevant features have themselves been written and marked by beacons, landmarks, boards, arrows, street names and so on. The easiest proof of this is to try to navigate with a very good map along an unmarked coast, or in a country where all the road boards have been torn off ... The chance is that you will soon be wrecked and lost. When the out-thereness is really encountered, when things out there are seen for the *first* time, this is the end of science, since the essential cause of scientific superiority has vanished. (p. 254)

Therefore, November, Camacho-Hübner, and Latour (2010) propose that we re-read both analogical and digital mapping in navigational terms (see Biemann, 2003). They want to draw attention, in particular, to the materiality of the production and consumption networks that resurface through the only apparent dematerialisation of the experience of digital navigation (p. 584). The idea that we have entered a whole new territory because of digital flows, as well as the attempt to delineate its new spatial features, is, indeed, very common, and is part of a longstanding belief in the autonomous existence of the *res extensa*. This 'res extensa effect', which could also be considered as a 'res imaginans effect' (p. 591) is nothing more than an imaginary construction relying on all the virtual images obtained from a mimetic interpretation of mapping techniques that, projected outside, create space as the most comprehensive virtual image which results in a complete erasure of material networks and of the technologies of representation.

To explain the difference between a mimetic and a navigational interpretation of maps, November, Camacho-Hübner, and Latour (2010) turn to the image of the navigator who, inside the cabin of a sailing ship, tries to combine both the signs on the paper map and the instructions of the crew that arrive from the deck above. Of course, the navigator never thinks that s/he inhabits a geometrical space and that the team members in the deck belong to the outside world. What s/he tries to understand is not some form of correspondence between the image of the world on the map and the real world outside, but the necessary cues that allow the crew 'to go through a heterogeneous set of datapoints from one signpost to the next' (p. 585). There exist many signposts that work at establishing a correspondence between the map and the territory, but at least two very different meanings can be attributed to the notion of correspondence at stake here: one relies on the *resemblance* between signs belonging to the map and things belonging to the world, whereas the other depends on the '*relevance* that allows a *navigator to align several successive signposts along a trajectory*' (p. 586). Whereas in the first case only a jump can bridge the huge gap between the order of representations and the order of things, in the second case a '*deambulation*' between different signposts, what the authors call 'the *navigational* interpretation of maps' as opposed to the mimetic one (p. 586). While mimetic correspondences are used to separate the inside and the outside, in a navigational framework, correspondences are seen in their continuity along an uninterrupted series. So, what the authors call a 'spurious referent' can only be produced out of a representation which, being isolated from the chain to which it belongs, is made to assume a mimetic function. The aestheticisation of maps, for example, contrary to what we might assume, does not concern the consideration of the artistic value of the map taken as an artefact, but the assimilation of maps to the chain of significations to which paintings belong: it derives from the belief that they work mimetically according to the 'one copy one model mode' (p. 590), despite the fact that the practical employment of maps continuously contradicts such a presupposition.

Navigational maps, on the contrary, cannot be simply looked at by a distant observer. They need to be experienced together with places by an observer that is placed in their midst. So, a distinction like that which de Certeau (1984) – who, in a sense, also prefigures non-representational conceptualisations of space – maintains between the itinerary as a discursive series of operations and the map as a totalising projection of ordered observations appears misleading in contemporary technospaces (see Bratton, 2013). Consider, for example, a smartphone application like the *MapFan Eye* (MapFanIncrement P Corp, 2012), a navigator whose AR interface does not turn the users away from their context but performs physical spaces in conjunction with the itineraries that are traced along/through them. It superimposes a transparent layer which augments what the users see, so that they can use the application without switching their view from the screen to the physical reality which they are in, but can follow their actual route through the screen itself. Here, maps and spaces interface in a continuous process of reciprocal mediation, generating an interactive topology very similar to what Vincent J. Del Casino, Jr. and Stephen P. Hanna (2006), drawing on the tradition of critical cartography and ANT, have called *map space*s, spaces where disjoining 'representations from performances' (p. 44) becomes impossible in the end.

3.4   A Politics of Location for Locative Media

Geolocalised information and communication today have become not only the content but also the context – pervasive and surrounding – of our interactions (Gordon and de Souza e Silva, 2011). The term 'geomedia'[5] (Thielmann, 2010), has been coined to name the current entanglements of media and geography. The end of space, which has been feared for so long and has commonly been attributed to the 'advent' of information society, is recast as a set of transformations that interest places and media at the same time, giving rise to media that appear to be increasingly located and localities that become more and more mediated (see Mitchell, W.J.T. 2008). In the geomedia environment, in which places look like 'palimpsests' (de Certeau 1984, p. 109), 'maps not only clarify what we already know; they present the conditions for reframing the questions', say Gordon and Adriana de Souza e Silva (2011, p.24). A sequence of overlapping cartographies brings attention to the dislocations that maps perform rather than to their approximation to an always ineffable origin as their source (see also Deleuze cited in Thrift, 2008, p. 118), contributing to the reframing of the issues of representation and spatiality in performative terms.

Besides opening up a different scenario for contemporary social actors – human and non – in terms of new affordances and constraints, the geospatial Web allows the two poles of this renewed debate – the presumed concreteness of location and the presumed abstraction of information – to be cast in a more complex, entangled scenario that leads us to consider both places and technologies of location as intertwined from the very beginning (see de Souza e Silva, 2006). Geomedia practices are characterised by a participatory relationality linking social actors through a diffuse addressability (Mitchell, W.J.T. 2008) in which corporeality, materiality and location are performatively redefined. The performativity of locative media practices, at the same time, appears as an embodied experience in a media environment which is more and more immersive and localised, and as the mapping of a material/informational spatiality which is itself performative.

Rather than dealing with new spatial forms or recovering old ones, such performances require a totally different conceptualisation of place. In fact, the geomediation of place makes different navigations of places available more than it alters 'traditional' places. Lemos (2008) has, for example, coined the term 'informational territory' to describe the way digital flows of information create new *functions* for the social practice of places, that is, new heterotopias in the Foucauldian sense (see Foucault, 1986), as he says, rather than new spatial *forms*.

---

5   The use of a term such as 'geomedia' denotes the impossibility of considering locative media and mediated localities separately, as well as the fact that the spatial turn of media studies has been accompanied by a substantial media turn in geography which has, for example, produced such neologisms as 'Neogeography'. This term defines the diffusion and employment of locational applications beyond the field of professional geography, so as 'to include experimental and artistic practices that come "from the outside"' (Haden, 2008; see also Cerda Seguel, 2009).

Accordingly, Rob Kitchin and Martin Dodge (2011) talk about code/spaces, which occur whenever software and spatiality are mutually co-constituted (p. 16). Such technospaces are less static (allowing for a conjoined consideration of location and mobility), as we have seen, and also less spatial in a traditional sense, since they possess an ontogenetic quality 'that lack[s] a secure ontology' (p. 16). Locative mobile media address technospaces that are continuously remade through practices and relations which are not intrinsically predetermined, but depend on contingent interactions.

Being both the context and the content for interaction, geolocalised information in fact requires engagement instead of contemplation. As de Souza e Silva and Daniel Soutko (2011) show, discussing the application *WikiMe* in which Wikipedia information is shown in relation to the user's position, physical location becomes the interface, or 'the marker', to access information. For example, users can access information on a certain museum only when they are physically near that museum. So, this application creates the situation in which a user might go to a place to *experience information* about a place, and in this way be able to access a hybrid dimension of informational place. This specific affordance of location-based applications turns the representational logic that regulates the link between signs and things, and the traditional employment of maps, upside down: physical space (as the signifier) becomes the sign from which to access informational space (as the signified). But the distinction is only an abstraction. As a matter of fact, 'The map of one's knowledge – *WikiMe* – precedes the territory the user navigates. However, that map of the territory is created dynamically through the interaction between user and physical space, and in turn the user interaction with the physical space is informed by and reflected on the map' (p. 30).

Technically speaking, locative media include the information and communication environments and practices that work through location-aware technologies, such as GPS, RFID, wireless networks, ubiquitous computers, smartphones and wearables, and that can be accessed from either mobile or fixed technologies – although the locational aggregation of data is only foundational for mobile applications (Gordon and de Souza e Silva, 2011, p. 11). In this context, user production and modifiability of digital maps, that some technical steps have made possible,[6] appeal to a positioned 'observer' who is increasingly summoned to 'access, alter and deploy networked data' through locative interfaces (Gordon and de Souza e Silva, 2011, p. 29). In fact, locative applications of digital technologies allow not only the use of information and communication in a localised way, but also the production of VGI (*Volunteered Geographical Information*) and active participation in GIS, often from within the same platforms where

---

6   An example of this is the availability of Google's API in 2005, Google's donation of the Keyhole Markup Language (KML, the necessary tool for the geoannotation and visualisation of digital maps) to the Open Geospatial Consortium in 2008, and the employment of Free and Open Source Software (FOSS).

tactical and strategic uses conflictingly coexist (Holmes, 2000; Crandall, 2006; Crampton, 2009; Kingsbury and Jones, 2009).

It must be noticed, however, that, apart from the novelty of the manifold digital applications that characterise locative tools, location-based media cannot only be considered under the lens of their novelty: firstly, because all media, including those for which distance and immateriality apparently play the dominant role, such as satellite technologies, are locatively produced and consumed, part of an ongoing process of negotiation and recombination; secondly, because even what goes under the name of locative media today must be considered in a relation of convergence with many other media and existing mediated and mediating practices (Bolter and Grusin, 1999; Poster, 2004; Hayles, 2005; Parks, 2005). In this respect, all media function as 'global positioning systems' (Meyrowitz, 2005, p. 24).

To further complicate the picture, while the technical definition of locative media clearly delineates a field of usage for locative devices, it does not automatically refer to a situated employment of such media. It may be useful, then, return to the definition of locative media that was initially elaborated in the art field, in order to extend it beyond a simple technical understanding. In fact, keeping in mind the conjoined genealogy of locative media and locative arts can help us better approach the performative qualities of the media environment without falling back on representational validation.

Artistic interventions employing locative media range from tagging, geoannotation and storytelling, which can be grouped under the umbrella of 'experiential mapping' (Bleecker and Knowlton, 2006), to wearables, games and theatrical events; in effect, all art has always dealt with location in varying degrees, as Hemment (2006) highlights in the first comprehensive essay on locative arts, starting from the relation of the artwork to its context of production and consumption. Some artistic movements more than others, then, have deployed an aesthetics of location rooted in a politics of situational engagement entailing different dynamics of situatedness and mobility: we need only think of the counter-cultural practices dating back to the late 1960s, often based on a performative dialectics between mobility and site-specificity, such as Performance Art, Arte Povera, Land Art and the Situationists' experiments.

The term 'locative media', coined by Karlis Kalnins during a workshop in Karosta, Latvia, in the Summer of 2003, broadly denotes artistic uses of location-based media as opposed to their corporate applications (see Crow et al., 2008). Actually, Kalnins notes, 'the locative case, in Finnish, roughly corresponds in English to the *preposition* "in", "at", or "by", indicating the types of proximity or relationality that we have to a given territory' (cited in Crow et al., 2008). Noticeably, here the term does not merely regard the technical possibilities of mapping and localising that are permitted by locative devices, but also the practice of places in their performative and embodied dimension. In the introductory essay of the *Transcultural Mapping Reader* (2006), Ben Russell, the author of *The Headmap Manifesto* (1999) – considered the first manifesto on locative media (see Tuters and Varnelis, 2006) – notes that locative technologies are not merely hard devices but also encompass a metaphorical dimension in which new places of thinking, seeing and doing are activated at the same time.

On the one hand, locative media indicate a new conceptual framework within which to discuss different epistemological approaches to hybrid places in their mediations with (not necessarily new) information and communication technologies; on the other hand, they delineate a new critical area in which the hegemonic uses of locative tools can be analysed in order to tactically utilise their possibilities for creative and user-oriented purposes (Russell, 2006). The latter is an aspect that has not always been pursued, especially since locative media arts easily find commercial funding and application, as for example the corporate-funded pioneering projects of the Proboscis art group, like the work of mobile annotation *Urban Tapestries* (Proboscis, 2002–2004), show (see Hemment, 2006; Townsend, 2006; Tuters and Varnelis 2006; Lemos 2008; Gordon and de Souza e Silva, 2011).

In theory, locative media arts are supposed to manifest a return of the digital to its historical and geographical embeddedness (Sassen, 2002; Hemment, 2004), contrary to many assumptions about the artistic autonomy of Net Art (see Tuters and Varnelis, 2006). However, somehow contradictorily, they also actualise the experiments with communication and information of many conceptual artists of the late 1960s and 1970s in which locative tools like maps and grids were often used to advocate for the objectivity of the work of art (Wollen, 1999; Fusco, 2004; Pope, 2005). So, when Drew Hemment (2004) proposes that we define locative media as 'embedded media', a definition that stresses the pervasiveness of media technologies in all experiences of space, he is also foregrounding the ambiguity of such experiences. The embeddedness of locative media is part and parcel of their complicity with the power to chart a territory for commercial and military purposes – to cite only two hegemonic uses of locative tools – employing the same locative networks and devices, from mobile phones and RFID to GPS and GIS.

Ultimately, just as locative technologies are not *stricto sensu* the precondition for a locative use of information and communication in technospaces, not all locative media projects necessarily rely on a situated approach. Locative tools often also highlight a dimension in which a traditional notion of subjectivity and spatiality is reinforced rather than dislocated, often by way of a fascination with the visual device and the abstract quality of locative objects such as maps. It is worth noting that, in fact, some of the weakest aspects of locative media arts resurface around the issue of cartographic representation. The unquestioned and reductive notions of objectivity and spatiality that they often presuppose leads to a distancing from embodiment and context, the latter existing only as residues of the coordinate system, and an insistence on a perfect correspondence between image and world, and a tendency towards the assumption of either utopian or dystopian approaches, are both distant from a situated politics of positioning (see Pope, 2005).

Exchanges between cartographic figurations and artistic practices long precede the appearance of locative media. If, on the one hand, Mercator was not only a cartographer but also an engraver and calligrapher, on the other hand, the interests of Leonardo da Vinci and Albrecht Dürer in cartography are documented (Rees, 1980;

Wood, 2006). Initially used in a decorative manner and then as objective descriptions of territories, maps have been largely employed in the visual arts to get rid of traditional forms of representation, a practice already common within Surrealism and, more frequently, after the second half of the 1950s (see Wood, 2006).

The rise of so called 'map art', in a period of the 1960s to the 1990s, has paralleled the increasing availability of maps in technospaces. The more maps have become ubiquitous and their descriptive function emphasised, the more the arts have tried to envision the creativity of maps beyond their normative meanings. As Denis Wood (2006) affirms, when contemporary artists had already employed maps in practices of countermapping, before the birth of locative arts, they intended to reject the authority of maps as a confirmation of the status quo and, rather than using them as 'descriptions *of the territory*', they employed them as 'descriptions *of the behaviors linked through the territory*' (p. 8).

Situationist psychogeography experiments with cartography have often been considered antecedents of locative media. Situationism was a European artistic and political movement active between the end of the 1950s and the beginning of the 1970s. The aim of psychogeography, a hybrid approach half play and half urban methodology, was to deconstruct the codes of a territory through the creation of new connections and disconnections between the physical and architectural structures and the rational or emotional representations that people can have of them. By means of the drift (*dérive*), defined by Guy Debord (1998) as a 'ludic-constructive behaviour that differs from the classical concepts of journey and walk in every respect' (p. 56, author's translation), space was experienced as fluid and discontinuous, until the pre-set marked boundaries, which progressively vanished through the drift, tended to disappear completely. According to the Situationist approach, the *détournement* happened each time existing spatial elements were displaced and recombined, and its aim was disalienating citizens (Kotányi and Vaneigem 1998, p. 76), so that they could go back to 'authentic life', free from the conditionings of the useful and the artificial.

Situationism is frequently recalled in locative media practices today (see McGarrigle, 2009). Two examples are Mark Shepard's *The Serendipitor* (2010–2011) and James Bridle's *Robot Flâneur* (2011). The former is a navigational application for the iPhone that leads the user along a path whose complexity can be decided beyond its mere functionality: users enter a destination, and on this basis the application proposes a route which combines the directions of existing routing services with artistic instructions aimed at creating detours within the route's *economy*, depending on the time the user can spend. *Robot Flâneur* is an instrument for exploring Google Street View which collects images from nine cities which automatically change every 30 seconds or can be switched by the user. However, the Situationist critique of society and space appears problematic by now, both for its utopic approach – seeking the revolutionary transformation of the everyday, the end of capitalism and the actualisation of art (Zeffiro, 2012) – and its dystopic one – struggling against the pervasivity of power and industrial ideology, which require a complete subversion. So, according to Andrea Zeffiro (2012), the critique of Situationism's influence on locative media should be twofold: on the one hand, it should foreground the shallow recovery of its revolutionary inspiration, and on the other, it should question the contemporary validity of the Situationist paradigm (p. 255).

Not only does the contemporary revival of Situationism in locative media practices still maintain a human-centric root, but it also continues to be linked to representational epistemology. After all, Situationism was against inauthentic representations of life (what would become the simulacra), but it believed in a rediscovery of authentic ones. Marc Tuters (2010; 2012) highlights what he sees as the weakest aspect – he calls it 'Mannerist Situationism' – of the contemporary drift in locative media: believing that a domain outside of capital still exists, and that, consequently, an oppositional politics is possible, when in fact locative media have to work inside the same networks that they can differently appropriate, but certainly never completely subvert. A-critically evoking notions like *flânerie* (see Pope, 2005; Zeffiro 2012) and believing in the immediacy and authenticity of spatial experience is decisively at odds with the pervasive mediation of contemporary technospaces, as well as with its heterogeneous assemblages and intrinsic contradictions (Tuters, 2012).

This is why, in this respect, a framework like that of ANT appears more suitable. A 'new ecological politics' (Tuters, 2010) composed of amateurish, distributed and collaborative actions from within the urban environment – what goes under the term 'citizen science'[7] – seems a more appropriate approach than the subverting tactics of Situationism to understand how human and non-human actors *gather*, in Latourian terms, or intra-act (see Barad, 2007), in contemporary technospaces. Adopting an ANT approach for locative media leaves room for the encounters *in the middle* (Mitchell, W.J.T. 2008) between what was thought to be 'out there'(see Latour, 1987), namely the objects (as things), and what remained inside, namely the subjects (as human beings). Above all, blurring the distinction between human and non-human actors inside multiply mediated and jointly performed technospaces avoids framing the theory and practice of locative media inside a representational perspective, still implicit in most actualisations of Situationism.

Locative media are not merely the media environments defined by locative technologies become mobile; in fact, technologies perceive only a stable sequence of points in space (see Pope, 2005), whereas locative media comprise the locative-mobile relations between these points, which involve human and non-human actors. This is why, as Lisa Parks (2005) contends, the expression 'global positioning' should not only stand for the ensemble of locative technologies that rely on geometrical coordinates, but also for the very act of mobile positioning that is required whenever a technology is contextualised 'between technological uses, knowledge practices, and social positionalities that stretch across national borders'

---

7   Of very recent invention, the term indicates forms of participatory science and scientific activities in which nonexperts and amateurs voluntarily and a-hierarchically cooperate with professional scientists in order to achieve community-oriented goals. See Chapter 4.

(p. 181). Thus, even locative technologies and locative practices need to be located and approached from a situated perspective because their recourse to location is much less immediately legible than would appear in the first instance.

Embeddedness, in this respect, also means that locative media can work differently from traditional mapping tools, going beyond a representational level. They can be made to work at a performative level, as tools for mobilising a given representation of a territory and thus dismantling several consequential illusions, such as the ideas that places pre-exist their representations, that an exact correspondence exists between a representation and the space represented, or that only one representation can be the objective one. Whenever locative media are used not for pinning down but for opening up the top-down approaches of conventional cartography, they can initiate different practices and representations of technospaces, as well as reappropriate and repurpose locative technologies (Hemment, 2004; Townsend, 2006).

When locative media actively interface with our experiences of places, they can produce '*patterns of difference*', to quote Haraway (1997, p. 268), which displace the identity of places and their faithful representations at the same time. Hemment (2006), for example, suggests that we speak of *dis-locative media* to go beyond the conventional geometry of the grid, which recalls the displacement of the same elsewhere that Haraway's (1997) diffractions mobilise. When space can be elsewhere, exceeding its givenness, it can also be otherwise. In this tension, the actuality of location, being processual and not static, also reveals its virtuality, foregrounding the creative quality of space in the aforementioned sense: for each practiced space-time, a series of other possible space-times exist and can be 'represented'.

The political efficacy of maps increases when they mobilise either existing representations of space or the existing meanings related to them, dismantling given routes and associations and imagining alternative paths that can be accessed and navigated by actors who are usually excluded from the picture. As Hemment (2004) puts it:

> a politics that is distinct to locative media – a politics of location – is not immediately apparent. Locative media proposes a form of dissent that is 'collectively constructive rather than oppositional' ... Locative Media's political moment might not be *despite* its complicity in mechanisms of domination but *because* of it, residing in the acceptance of the paradox and occupying the ambiguous space it creates, creating a site of resistance by working from the inside.

The Bangladeshi-born American artist Hasan M. Elahi, detained for nine hours at the Detroit airport on the suspicion of stockpiling explosives and investigated by the FBI for suspected terrorism based solely on his ethnic origins, turns police control into a lifelogging performance working in a paradoxical space where surveillance mechanisms are turned upside down. His project, which is also a website initiated in 2006 and still ongoing, is entitled *Tracking Transience* and explores the border 'between society and technology [and] the intersection of geopolitical conditions and individual circumstances' (Elahi, n.d.). Here, Elahi records every insignificant aspect of his life – from meals to credit card transactions – by means of a device similar to the prison ankle bracelet. Each piece of information, which is also visually supported by many pictures, is geotagged and then uploaded to a server that sends the tag to the United States Geological Survey, which returns it in the form of an aerial surveillance image (visible at different scales on the project's web site). In this way, the artist not only pre-empts control by exposing himself to absolute visibility, but also empties visibility from its hegemonic significance, transforming images into performative gestures.

One of the clearest examples of performative locative media that works from a position of 'located accountability … situating the terms of access to militarist infrastructures and capitalist ventures' so as to 'disrupt rather than secure the act of appropriation' (Zeffiro, 2006) is *The Transborder Immigrant Tool* – TBT – (EDT, 2007–) by Ricardo Dominguez, along with other members of the Electronic Disturbance Theater (EDT) group, Brett Stalbaum, Micha Cárdenas, Jason Najarro and Amy Sara Carroll (b.a.n.g. lab and EDT 2.0., 2013).[8] Activist interventions of the EDT include, among other things, *Electronic Civil Disobedience* (ECD),[9] a practice by which the collective transforms the Net into a site of political action rather than using it merely as a communicative tool. In reality, the TBT is a mobile phone equipped with a GPS receiver and a specifically designed piece of software, developed at University of California, San Diego, by the EDT group.

Whereas Border Patrol agents, thanks to the BlueServo.net program, can rely on volunteers all over the world watching surveillance cameras set along the border to detect the movements of migrants and report them to the police, at the same time and thanks to the same locative technologies, migrants can attempt to trespass the border safely thanks to the TBT. The TBT, working like a compass, helps migrants orient themselves among several aid stations so as to safely trespass the USA-

---

8   Another interesting example of 'dislocative' artwork working on mobile devices and wireless networks at the border is *Border Bumping* (2012–2013) by Julian Oliver, a free downloadable application for smartphones which registers the trespassing 'incidents' of mobile devices at the borders, each time they traverse territorial limits before the owners do (Timeto, 2015).

9   ECD works through a simple application which allows the participants to set up an automated reload request every few seconds against the home page of the institution under attack; it reaches its utmost efficiency in a short period of time, creating an informational gap which renders unstable the institution that is the object of attack, thus allowing a momentary reconfiguration of power relationships. The inaugural Floodnet took place on 10 April 1998 following the Acteal (Chiapas) massacre of 1997 in which 45 indigenous people were killed. It was directed against Mexican President Zedillo and in support of the Zapatistas, who were already using the Net for their fight. Many other virtual sit-ins have been organised by the EDT since, including one protesting the ongoing femicide in Ciudad Juàrez (August 2002) and a more recent one (March 2008) against the Nano/Bio War Profiteers, on the occasion of the fifth anniversary of the Iraq war.

Figure 3.2   Photo of *The Transborder Immigrant Tool*, by Electronic Disturbance Theater, 2007

Source: Courtesy of Electronic Disturbance Theater 2.0./b.a.n.g. lab.

Mexico border without being detected. In the words of the artists, it is intended to help those who are usually excluded from what they call the '*emerging grid of hyper-geo-mapping-power*' to acquire '*situational awareness*' (EDT, 2009). This kind of operation, however, has in fact become more and more risky after NAFTA in 1994. In between walking art and locative media art, the TBT foregrounds the differences between and the entwining of bodies and data, material and immaterial flows, artistic wandering and migratory mobility, focussing on the issues of life and death that are involved in the traversing of borders. The artists draw on the idea of *performative technology* as it is conceived by the locative media artist Christian Nold (2009),[10] as a device used to mediate interpersonal relationships and build a sense of local community, which Brett Stalbaum (2006), the TBT software developer, echoes in his notion of *paradigmatic performance*. Stalbaum uses this term to define a technologically-based artistic practice that is not only collectively conceived but also employed for collective purposes, being '*generative of new configurations of practice*'.

In this respect, the performativity of the TBT is intended as both an aesthetic piece and an ethico-political intervention that is markedly political, since it is intended to directly improve people's lives, according to what Dominguez and his EDT collaborators identify as a shift from tactical media to *tactical biopolitics*

---

10   For his *Bio Mapping* (Nold, 2004–) project, participants wear 'galvanic' devices that record their emotional arousal in relation to the places they walk in, so that communal emotion maps can be produced.

(see also Zeffiro, 2015). Stalbaum has projected a *Virtual Hiker Algorithm* for it, so that the cell phone can be used as a GPS walking tool to mediate the actual experience of migrants crossing the border. But GPS, here, does not only stand for Global Positioning System; it is also an acronym for Global Poetic System. In fact, the cell phone is also endowed with a set of bilingual poems written by Amy Carroll that starts to play while walking: they are intended to psychologically assist migrants during their crossing and welcome them into a new space of hospitality and solidarity.

Made in collaboration with local collectives like the Border Angels, the project, for now at prototype stage, comprises several steps: from the GPS mapping of the coordinates of the border territory, as well as of support networks and anchor points, such as water and food stations, and the development of the specific software and bilingual (English/Spanish) interface, to the final distribution of the mobile phones to migrants on both sides of the border, who are supposed to return them for further use after reaching their final anchor point. In 2011, on the occasion of the cross-border event *Political Equator 3* that took place at the San Diego/Tjuana border in June, the b.a.n.g. lab (the EDT research lab at Calit2, an acronym for Bits, Atoms, Neurons, and Genes) tested the TBT: the phone was symbolically walked by artist Marlène Ramírez-Cancio into Mexico via a tunnel from the US side of the border (Dominguez, 2013).

The TBT has not only been the target of conservative media hype; as we can see from the press review collected on the b.a.n.g. lab and EDT 2.0 blog (2013–), it has also been an object of investigation, paralleling the investigation by the Audit and Management Advisory Services at the University of California (UC), San Diego, that followed the virtual sit-in performance of 4 March 2010, promoted by the EDT. The virtual sit-in was launched against the homepage of UC President Mark Yudof in conjunction with UC-wide demonstrations against financial cuts as well as the racist atmosphere of the UC campuses. The virtual strike was accused of violating University policy and thus investigated which also caused an investigation of Dominguez, putting his tenure-track academic position at risk. The most important underlying reason was to find out whether the University funds employed for the TBT project were being used according to the motivations of their allocation or in violation of it. A Legal Action Fund was soon created to give economic support to Dominguez, but the charges against him were ultimately dropped the following July with the finding that the funds were being appropriately used.

This episode testifies to how borders operate at various levels and not only refer to territorial realities but also invest discursive networks, producing contested divisions that can become sites of struggle once their genealogy is reconstructed (see Mezzadra and Neilson, 2013). In this respect, the border itself should be seen as an epistemic device that is also productive, rather than merely descriptive, a performative representation for investigating the existent constitutions of subjects and spaces as well as for envisioning and practising alternative technospatial constitutions. We could say that the performativity of the border acts metonymically with regard to the performativity of space and its mapping.

The activity of the EDT brings to the fore the problem of the mobilisation of existing borders at various levels. Both the TBT and the practice of ECD, in fact, make transparent the shifting boundaries between theory and practice, art and politics, the material and the symbolic. The tactical approach of this work (EDT, 2009; Kroker, 2010) also refers to Chela Sandoval's book *Methodology of the Oppressed* (2000) as a source of inspiration. Her notion of differential consciousness, intended as a performative medium that activates a new tactical space for oppositional praxis (pp. 57–63), which she elaborates in the context of her feminist thought, can be compared to that of performative media in the sense used here. Sandoval believes that resistance is only effective when it is differently related to the diverse forms that power can assume, that is, when it is activated through specific tactics which vary contextually. The differential crosses the multiple networks of power, devising them also as tactical tools and thus allowing for a constant rearranging of boundaries (p. 181). Sandoval's is a situated methodology of positioning in that it refuses any oppositional based on rigid dichotomies that would preclude the crossing of boundaries, which alone allows for relationality and encounters in material and symbolic spaces.

If a representational map usually renders a route as a series of visible points that transform '*action into legibility*', to use de Certeau's words (1984, p. 97), a performative map is supposed to do exactly the opposite, re-embodying readable lines into active practices. Like a linguistic enunciation, walking presupposes a series of differential relations among people sharing the same space, says de Certeau:

> walking affirms, suspects, tries out, transgresses, respects, etc. the trajectories it 'speaks'. All the modalities sing a part in this chorus, changing from step to step, stepping in through proportions, sequences and intensities which vary according to the time, the path taken and the walker. These enunciatory operations are of an unlimited diversity. They therefore cannot be reduced to their graphic trail. (p. 99)

Accordingly, the TBT works performatively insofar as it accompanies the migrant along a journey which dislocates the corporate territorialisation of the border zone and performs *bottom-up* tactical functions of the same territory. Overwriting a zone of restricted mobility and surveillance networks like the USA-Mexico border with the routes of migrants that are put in the condition of actually and safely cutting across such contested and regulated technospace, the TBT enacts a politics of location that foregrounds the performativity of locative media in conjunction with the performativity of space. It shows how digital space cannot be fully experienced unless it is embedded in the material conditions allowing for its existence and imagination.

Chapter 4
Diffracting Technoscience

4.1 The Open Lab

The dislocation that works like the TBT (see Chapter 3) effectuate happens on a horizontal, anti-hierarchical level, so to speak. It challenges the transcendentalism of the tradition of spatial exploration which permeates much of the contemporary hype about locative mobile technologies and replaces it with issues of bare life, emergency and hospitality – in a word, with technobiopolitics. It is no coincidence that Dominguez is also a co-founder of the *particle group*, a collective of artists and activists, working since 2006, that adopts a multimedia approach to investigate the spaces of tactical resistance inside the corporate networks of nanotechnologies.

As Shields (2006) writes, 'everyday sites need to be re-thought as milieux interlaced with political and biotechnical processes happening at nano-scale' (p. 217). These are the technospaces in which contemporary cyborgs live. Split and hybrid figures, cyborgs actually move inside 'a virtual terrain of struggle' comprising spaces 'of nano- and biotechnology *beneath* the scales at which domination has been understood to operate socially and politically' (p. 218). Given that cyborgs exceed the boundaries of the body, it is reductive to consider them as sites of inscription of biopowers in the sense Foucault (1979) intends it, says Shields (see also Barad, 2007, pp. 200 ff.). Indeed, expanding on Foucault's analysis, Haraway (1997) prefers using the term technobiopower to analyse the dynamics of the production and reproduction of cyborg bodies inside multiple global flows.

Foucault stressed the immanent forces, both regulative and productive, that create a coimplication between the discourses of sexuality and the power relations at stake in specific forms of knowledge concerning bodies. Haraway (1997) further writes: 'All the entities in technoscience are constituted *in* the action of knowledge production, not before the action starts'(p. 29), where they co-emerge '*through the constitutive practices of technoscience production themselves*' (p. 35; see also Wajcman, 2004). This statement implies two important consequences: blurring the boundaries between the subject and the object of technoscience, it evidences that bodies are always materially as well as symbolically *in the making* and that science, rather than being an approximation, discovery and description of essential truth, also engages in contingent practices and partial truths (see Barad, 2003). The contemporary spatialisation of the cyborg urges us to reconsider traditional sociospatial categories still based on the representation of homogenous, unitary social actors at all scales, from the individual subject to the nation. The interfacial landscape of cyborgs 'operates not at the material level of the body but as a fractal body (e.g. regrowing organs and replacing body parts), at a nanotechnical scale (e.g. manipulating stem cells), with impacts which reverberate up in spatial scale and out temporally as a signal which changes the surrounding milieu' (Shields, 2006, p. 217).

In its work, subRosa addresses the 'distributed' and fragmented bodies that populate contemporary technospaces, which cannot be identified along singular coordinates but manifest simultaneously as the 'medicalized body, socially networked body, cyborg body, citizen body, virtual body, laboring body, soldier body, animal body and gestating body' (subRosa, 2011, p. 16). subRosa[1] is a cyberfeminist collective based in the USA whose activity dates back to the late 1990s,[2] when Faith Wilding established a study group on sex and gender in the biotech era at Carnegie Mellon University. The aim of the study group was to discuss the production and circulation of texts and images on this topic and also to discover the links between feminist art and the new fields of bioart, new media art and the art-science exchanges (subRosa, 2011).

Faith Wilding had been one of the founders of the first Feminist Art Program at CalArts and a leading artist of the *Womanhouse*[3] (Chicago et al., 1972) project. Inside *Womanhouse*, she had created the *Crocheted Environment*, also known as the *Womb Room*, a sheltering tent for self-healing that was the result of collaborative work represented by crocheting, traditionally considered a feminine craft. This work also visualised the feminist creation of connections through weaving (see Plant, 1995), to which today's feminist practices of networking in the 'integrated circuit' can be traced back (Haraway, 1991, p. 170).

An integrated circuit – an expression which Haraway (1991, p. 165) borrows from Rachel Grossman (1980) – is a web of medical, military, labour and informational power forces where women and other subaltern subjects – including animals and plants – are valued and exchanged as commodities. Here, the hybridity of the cyborg body does not only appear as an empowering condition, but also as the result of several overlapping powers and their unequal effects. In this layered, scattered space, the feminist location of knowledge works relationally inside a multiple web of interconnections. subRosa's methods, such as conviviality, collective action, an enlarged version of consciousness-raising through panelling, networking and leafleting, are not very far removed from those of many feminist

---

1   For detailed visual and textual documentation of subRosa's projects, including links to websites for specific artworks, see the group website (subRosa, 2010) and the video-collection of selected projects (2005a).

2   The official birth-act of the group, however, is considered its first public intervention at the Next Five Minutes Festival in Amsterdam in 1999.

3   Judy Chicago, Miriam Schapiro and Wilding, together with a group of female students who took part in the first Feminist Art Program at CalArts, collaboratively worked to restore an abandoned house in Hollywood, L.A., which, starting in the autumn of 1971, was the theatre of one of the largest scale feminist installations ever mounted (Jones, 1996). For about three months, artists and students worked together at restoring and recreating each space of the house that took the name *Womanhouse*, assigning a specific feminist theme to each room.

artists of the 1970s. But the collective also reinterprets and actualises the feminist tactics of 'weaving', extending them to the digital Web as an interwoven space of consciousness raising, connectivity and political advocacy. Combining high- and low-tech media, its offline activities parallel its online ones in the form of webworks and documentation websites, which in turn function as a locus for further reflection and action.

subRosa also reframes the tradition of grassroots politics and feminist politics of location according to a transnational perspective (Fernandez and Wilding, 2002): as a group of feminist artists working *with* and *on* digital and bio-technologies, it foregrounds the embeddedness of scientific practices as well as the situatedness of women's lives inside the material-informational networks of global technobiopower. An association of activists with a medical focus, Women on Waves (WoW) adopts a similar approach. WoW is a non-profit organisation founded by the Dutch gynaecologist and feminist activist Rebecca Gomperts in 1999. It basically operates on a ship that sails to countries where abortion is illegal, and provides contraceptives and safe medical abortions in transnational waters, as well as sexual education and advocacy. The organisation also works via a website (WoW, n.d.) offering counseling and the possibility of sharing one's experiences. WoW employs a variety of media to launch and promote its campaigns, including the visual arts. For example, the mobile clinic of the WoW ship stands inside the *A-Portable* container, a comforting and functional space designed by the artist Joep Van Lieshout, founder of the Atelier Van Lieshout. *A-Portable*, which was also exhibited at the 49th Venice Biennale in 2001, materialises Gomperts and Van Lieshout's idea that the aesthetic and the social cannot be easily disjoined.

Like subRosa, WoW focusses on the distributed bodies of the contemporary integrated circuit, acting at multiple spatial levels and with different media to push the boundaries of institutional closed fields and to create interstitial zones of action. WoW's and subRosa's activities can be compared in many respects. *Can You See Us Now?* (subRosa, 2004b), for example, is an installation in which subRosa maps the intersections of women's affective and material labour in the cultures of production of the former manufacturing and mill town of North Adams (MA) and Ciudad Juárez, Mexico, the same border town on which Biemann's video essay *Performing the Border* focusses (see Chapter 3). Tracking the history of North Adams's Sprague Electric factory, which boomed with the production of capacitors for civilian and military use during World War II, subRosa follows the transformation of a family business gone global: actually, most of Sprague Electric's manufacturing has been relocated to Juárez, where the *maquiladora* industry has grown drastically since the NAFTA agreement in 1994.

The effects of the global economy on affective labour, manual work, the service and care industries, the tourism economy, health conditions and reproductive technologies are all randomly mapped onto the walls and the 'forensic' floor of the subRosa installation, and need to be connected by the visitors. At the same time, this mapping allows another history to emerge: one of solidarity, struggle and resistance, in which autonomous zones such as education and support centres attempt to build an alternative – unfortunately equally silenced – history of the two towns. At the entrance, the installation also includes a map where visitors can pin the label of their clothes to visualise the trajectories of objects in the garment industries and the way these intersect with women's mobility.

Like *Can You See Us Now?*, the action that WoW launched in February 2012, although primarily promoting the use of Misoprostol – a synthetic prostaglandin used for a safe medical, that is non-surgical, abortion – also addresses the condition of women in the integrated circuit, specifically focussing on the garment industry (WoW, 2012a). The target of the campaign was the fashion brand Diesel, chosen as an example of an exploiter of women workers living in unhealthy conditions and earning a wage far below the legal minimum, particularly in developing countries. WoW's fake press release and the fake ads imitate Diesel's glossy style, using supermodels in rarefied ultra-tech settings, inviting the viewer to visit the campaign website (which is a mirror site,[4] along the lines of artivist groups like The Yes Men, who incidentally took part in its creation). The fake press release (WoW, 2012b) reads as follows:

> After launching Diesel Island, Land of the Stupid and Home of the Brave,[5] Diesel now creates Misopolis, a factory where brave female workers can have happy accidents without consequences. Misopolis will be the least fucked-up fashion factory in the world. But this is not just another factory – it is a destination that finally grants them real autonomy.

In one of the spoof ads created by WoW we read, for example, 'Say goodbye to coat hangers', which although used to hang clothes are also infamously known as abortion tools. A group of women stand around a table, where another woman partially lies with a blood stained t-shirt, and are holding a bunch of coat hangers, while the barcode on their tees, if scanned, offers information about Misoprostol. Everything is clearly extremely 'staged', but since this is the tone of the Diesel pictures as well, the difference is hard to recognise at first glance. Another ad shows an altar where a woman in a futuristic golden outfit (The immaculate 'Contraception') feeds one girl an abortion pill rather than the Host. And on each of these images, the actual Diesel slogan is repeated with only a slight difference, so that we now read 'Abortions for successful living'.

Reality, however, is very different from the cheerful and glossy one depicted here. As we read in the WoW press release, following the letter from Diesel that threated to take legal action, the hoax 'intends to show that violations of human rights never happen in isolation and that the right to a safe abortion is connected with the broader framework of social rights, workers' rights and the right to autonomy'

---

4   Technically, a mirror site replicates the original one but has a different URL. However, in art practices, mirror sites are often used to deface original content, while maintaining the same visual design.

5   These are all names of real, sexist campaigns by Diesel.

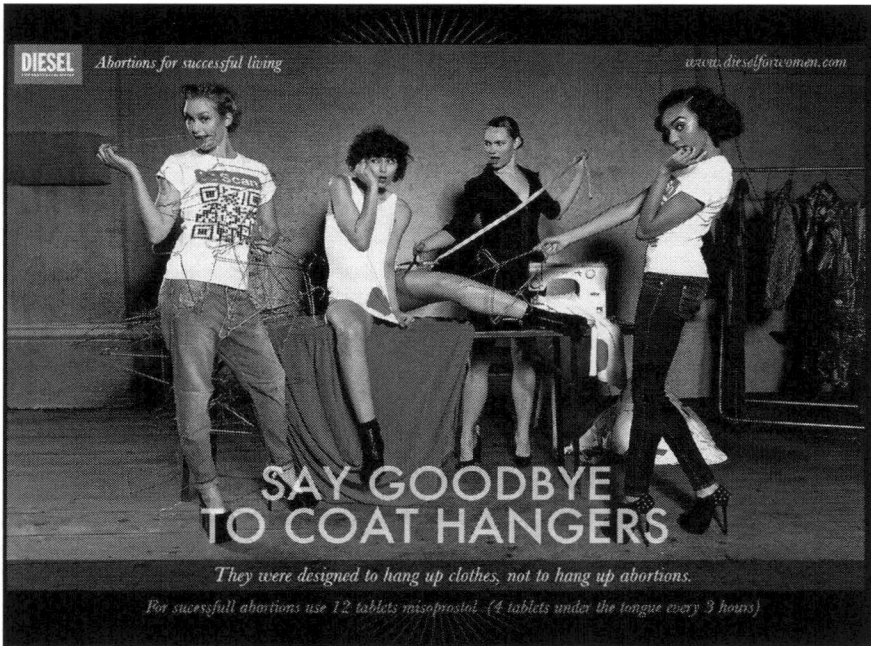

Figure 4.1   Advertisment for the *Misoprostol Campaign*, by Women on Waves, 2012

Source: Courtesy of Rebecca Gomperts.

(WoW, 2012c). In fact, between 75 per cent and 90 per cent of garment industry workers are women, very often young and uneducated, forced to work for many hours without a contract, and subject to sexual harassment, rape, and consequently unwanted pregnancies without, however, any right to maternity leave. Things get even more complicated because not only do female workers fear being fired if they claim their rights, they also often live in countries where abortion is illegal.

To make its interventions even more effective on a material level, subRosa usually accompanies its multimedia installations with situated performative actions. As a way to escape the visual essentialisation of the feminine without being relegated to invisibility, and to contest the exclusivity of the established canons of the art world, performance art and performativity have been at the core of feminist art and activism from the very beginning. Performance art inaugurates 'another representational economy', one that does without representation (Schneider, 1997, p. 3), transforming the work of art from a fetish into an operational event. Performances are contingent and ephemeral, virtually repeatable but nonetheless always differently situated. The first feminist performers experimented with the provisional, un-fixity, deformity and the formless, in order to contrast the fantasies that circumvented the feminine body and the work of art, accompanied by the desire to possess them both. Pregnancy, menstruation, dieting, ageing, surgeries, pleasure, violence, rape – everything that shows the passage from integrity to fragmentation, from the closed organism to the liminal body, was explored by feminist visual performers.

In subRosa's cyberfeminist interventions, performativity acquires renewed importance as a means to engage with technoscience. In fact, like many of the artists working with bioart today (see Hauser, 2008), subRosa's works *perform* scientific knowledge's performativity, so to speak, bypassing the representation of the objects of art and science to stage the way they are, instead, *made* in both domains (see Latour, 1987; 2010). The 'site-u-ational' approach of the collective, as they call it (subRosa, 2004a) – which finds analogies in the modes and scope of the 'recombinant theatre' of the Critical Art Ensemble (CAE, 2000) – aims at involving the audience in a public debate on such themes so as to counter the private theatre of technoscience (subRosa, 2003a) and to realign knowledge-making practices and artistic creativity.

Science is a recombinant, creative practice, and so are our technobodies: scientific knowledge and scientific objects, as they are enacted and shared in subRosa's performances, are the contingent result of thick, located experiences, which consist of interactions and exchanges among all the participants where the artists do not retain any privileged role but simply 'fabricate' their mediated position together with the audience's. The open lab in which subRosa performs technoscience practices is 'a place for the collective craft work of knowledge-making' (Haraway, 1997, p. 66) in which the artists's role is to restage and witness the making of science that, as art making, draws on a combination of 'cultural practice and practical culture' (p. 66). Significantly, subRosa often uses cut and paste and DIY techniques in its participatory performances, such as *Epidermic! DIY Cell Lab* (subRosa, 2005b), in which the audience, with the help of the collective's members doing 'bench-side work' (subRosa, 2011), is taught how to streak a Petri dish and how to make yogurt, with the aim of demystifying the myth of science and its 'alchemical imagery'. In subRosa's multimodal environments, people have the opportunity to learn by reading texts, watching videos and even eating themed snacks that the artists jokingly distribute.

For the exhibition *Knowing Bodies* (subRosa, 2000a), for example, subRosa puts together three interconnected pieces drawing on vaginal iconology and the maternal, as in the tradition of feminist art: a giant soft sculpture reproducing a vagina, which the audience is allowed to construct and manipulate; a video-performance, *Vulva De/Reconstructa*, about aesthetic surgery on female genitalia; and a webwork, *Smart Mom* (subRosa, 2009), about the possibility (passed off as plausible) of monitoring the pregnant mother and the foetus via remote control sensor-equipped suits. The re-appropriation of the feminine dimension, however, is problematised as soon as subRosa foregrounds the implications of such bodily enhancements: the audience learns that there exist many different – not only aesthetic – reasons for vulvar surgery, for instance genital mutilation, and also that very often the request for 'vaginal rejuvenation', behind the promise of a renewal of sexual pleasure, disguises the pressure to conform to the heterosexual and patriarchal norm (Wilding, 2002). 'Smart' technologies, too, are traced back to their military origin and their possible application to the control and normalisation of deviant bodies.

Figure 4.2 **Photo of the performance *Epidermic! DIY Cell Lab* (foreground) and of the installation *Cell Track: Mapping the Appropriation of Life Materials* by subRosa at *Soft Power. Art and Technologies in the Biopolitical Age* (background), curated by Maria Ptqk, Amarika Project at Vitoria-Gasteiz, Spain. October 30, 2009**

Source: Courtesy of subRosa.

Typically, subRosa unmasks the theatre of technoscience by re-employing and displacing its power from the inside through mimicry. A performative tactics par excellence, and for this reason often employed in performance art, mimicry short-circuits the logic of reflection of classical representation, introducing a disturbing gap that interferes with the resemblance to an original (see Butler, 1990). Like other feminist performers before, such as Hannah Wilke, Adrian Piper, Eleanor Antin, Suzanne Lacy, to cite only a few, but also like some contemporary artivists such as the Yes Men and the CAE, subRosa members act 'as if' (Braidotti, 1994, p. 7), to devise the intervals of power among repetitions, where 'alternative forms of agency' are made possible.

In the participatory art of subRosa, the performances of technobiopower are restaged and mimicked to reveal their linkages with the performances of everyday life. Works such as *Sex and Gender in the Biotech Century* (subRosa, 2000b) and *Expo Emmagenics* (subRosa, 2001a) adopt the strategies of corporations and turn them upside down. In the former, people take part in a fake class held by subRosa members posing as corporate and government delegates: while compiling a sort of exercise book and learning about Assisted Reproductive Technologies (A.R.T.), the participants in fact learn how bodies are accorded a different market value on the basis of ethnic, class and geographical factors. For *Expo Emmagenics*, subRosa's members pretend to be representatives of leading US firms preparing a trade show targeted toward the European market, where the latest American products related to A.R.T. are promoted: lively demonstrations explain how to use the *MegaBytes Tasties* and *Human Caviar* – resulting from the excess eggs of hormonal stimulation – as fertility and sexual revitalisers, as well as sperm saver condoms, do-it-yourself kits for in vitro fertilisation and GPS devices to find the perfect mate for 'producing' the ideal child.

Whereas the narratives of A.R.T. are often based on the rhetoric of choice and the manipulation of desire conveyed through the neutral and normalising language of technoscience, subRosa's performative mimicry discloses very different accounts which are class, race and gender-targeted. At issue are the ways women, notwithstanding appeals to individual freedom, are still addressed as objects of investigation and consumption, their bodies treated either as laboratories or resources, according to uneven routes of mobility that very often have colonial and eugenic implications (subRosa, 2002): consider, for instance, the similarities between the illegal traffic in organs and the legal mobility of egg and sperm cells, explored by subRosa in *International Markets of Flesh* (subRosa, 2003b), or the growing patenting of stem cells and seeds as a way to manage diversity and privatise common resources, at the centre of *Cell Track* (subRosa, 2004c) and *Epidermic! DIY Cell Lab* (subRosa, 2005b).

Cell Track is an installation and a website investigating the privatisation of human, plant and animal genomes. In the installation, a body combining male and female parts is mapped with a dymaxion map and horizontally bisected by a timeline in which important moments in the history of patenting are pinpointed. The website (subRosa, 2004c) also offers a great number of source materials, including a booklet that can be downloaded for free (*Cultures of Eugenics*), a glossary, didactic animations, and also the *Manifesto for a Post-Genome World*, which suggests that a 'democratic, creative and beneficial use of genetic knowledge', in which difference is evaluated but not fetishised and responsibility is equally distributed, is still possible.

Retracing the tradition of situated epistemology and situated knowledge, subRosa acknowledges that, to use Haraway's (1991) words,

> because science is part of the process of realizing and elaborating our own
> nature, of constituting the category of nature in the first place, our responsibility

for a feminist and socialist science is complex. We are far from understanding precisely what our biology might be, but we are beginning to know that its promise is rooted in our actual lives, that we have the science we make historically. (p. 45)

This also implies dealing with technobiopowers from within their sites and networks, 'refusing an anti-science metaphysics, a demonology of technology' (p. 181). Thus, like the WoW ship that *makes waves* on the homogeneous surface of striated space (see Deleuze and Guattari, 1987), subRosa creates spatial pockets of resistance, 'becoming autonomous zones'[6] (subRosa, 2001b) that it calls Refugia. Refugia should be conceived as contingent shelters rather than stable dwellings: they are heterogeneous spaces of encounter, recombinant technospaces against every form of monoculture. Refugia do not work as containers, since they do not simply collect but also disseminate, working instead as dissipative systems with porous boundaries. Situated, adaptable and reproducible, Refugia are unregulated zones that interfere with the highly regulated spaces of technobiopower and whose ephemerality is conceived specifically to resist corporate control. They also defy traditional spatial and temporal logics: neither utopias nor dystopias, they are slow-down situational spaces of 'imaginative inertia' (subRosa, 2001b), affect and desire. In fact, Refugia can also be useless and playful, while they nonetheless generate shared knowledge and cues for responsible action.

As subRosa's recombining of art and science practices demonstrate, a non-innocent (Haraway, 1991; 1997) but still factually valid approach to reality-making exists. Latour, for example, calls it 'second empiricism' (2004) or 'mediated' empiricism (2005), and Lorraine Code (2006) terms it 'negotiated empiricism': such approaches reconcile facts with criticism, while declaring the location of that which contributes to constituting knowing processes as they emerge. Actually, as Haraway (1997) puts it, 'both the subjects and objects of technoscience are forged and branded in the crucible of specific, located practices, some of which are global in their location' (p. 35). A different realist attitude looks undialectically at facts and objects as 'gatherings' (Latour, 2005, p. 115) and 'associations' (p. 119), that is, assemblages with a traceable genealogy[7] extending along different space-times. Thus, a critical approach does not stand as the opposite of a realist one, insofar as criticising what Latour calls 'matters of facts' brings criticism closer to facts rather than farther, 'towards the conditions that made them possible' (Latour, 2004, p. 231). A critical exercise reverses the natural consistency in which facts appear and reveals that they are but solidified 'matters of concern', which include beliefs,

---

6   This is a term drawing on Hakim Bey's (1985) T.A.Z, or 'temporary autonomous zones'.

7   'The term *fact,* with its inbuilt finality, fixity, obliterates genealogical traces, erases marks of the interactive, often conflictual labor out of which ... "factors" successfully (if for now) promote some candidates to the status of factuality, suppressing others as they go' (Code, 2006, pp. 101–2; see also Latour, 2010).

opinions, values, controversies. Criticism *diffracts* given realities 'cultivating "turning points"' for 'bringing apparatuses[8] to crisis' (Anderson, 2014, p. 19).

The trespassing of the enclosed and self-consistent spaces of scientific knowledge production that subRosa unlocks in its performances and that WoW makes navigable/cuts through in its actions, reveals the situatedness as well as the partiality of scientific practices, as well as 'their interworkings, their negotiated, deliberative character, and their social-environmental implications' (Code, 2006, p. 90). Scientific facts are not dismissed or devalued, but are situationally mediated by means of the crafting, creative, and imaginative means of art making. This also includes disclosing the practices through which situations are produced, that is reworking the situatedness of situations from the inside, since no site of knowledge production comes into being prior to the intra-active enacting of its actors and apparatuses (see Barad, 2003; 2007; Code, 2006).

4.2   An Ecosystemic Cure

As already foregrounded by Foucault in *The Birth of the Clinic* (2003), the understanding and conceptualisation of diseases and, accordingly, of cures, change over time depending on the organisation of modes of knowledge and objects to be known, which take place in spatial configurations and reciprocally arranged discourses. As for other fields of knowledge, the historicity of medicine defines its conditions of possibility, its domain and its rationale at the same time (p. xv). A series of visualising practices as well as representational analogies have been mobilised in the spatialisation of medical knowledge. In fact, as Foucault observes, in medicine, until the end of the eighteenth century – which for Foucault corresponds to a form of 'primary spatialization'– the abstraction of the individual patient's symptoms to arrive at the rule resembled the surface of a painting without depth, that worked as both the origin and the end of a process of recognition in which every disease was already classified in a table to which it should always be referred back (p. 9): working by way of analogy and resemblance, a 'map' [*repérage*] served to situate a symptom in a disease, and a disease, in turn, in a larger group that constituted a general, mathematical classification of pathologies (p. 39).

Later, at the end of the eighteenth century, when the medical gaze distanced itself from the clarity of an ideal mind that only guaranteed the exactitude of seeing and went out towards opaque bodies to become a more empirical exercise, the metaphor of painting started working in reverse: 'the patient [became] the rediscovered portrait of the disease; he [was] the disease itself, with shadow and relief, modulations, nuances, depth; and when describing the disease the doctor must strive to restore this living density' (Foucault, 2003, p. 15). It was as if, we could say, we had gone from the drawing to the texture of the same portrait,

---

8   On the concept of apparatus, see Barad (2007).

which eventually led medicine to 'carving up' (p. 29) the intersecting series of existing information with the series of observable variables. When the gaze dived into the deep, vertical dimension of pathological anatomy, another artistic metaphor resurfaced: the doctor was compared to an artist for sensibility and taste, and medical analysis was deployed at an aesthetic, plurisensorial level under the domain of sight as well (pp. 120 ff.). This move corresponds to what Foucault identifies as 'secondary spatialization', a formal reorganisation of the place of objectivity in the background of the tridimensional body (first the individual, then the collective), rather than in the background of the disembodied mind.

At the same time that the medical gaze explored corporeal depth, equivalences and homologies were looked for in a 'a ceaselessly supervised environment' (Foucault, 2003, p. 32) which became the 'free field' (p. 38) of a gaze which did not meet any obstacle in its exploratory surveys, but could float everywhere. This elevated the medical gaze to a quasi-religious status, transforming medicine into a social and political task in which ideologies and technologies converged in their common search for an identification, and possible elimination, of disease from the social fabric. 'Tertiary spatialization' was the spatial restructuring – in hospitals, faculties, medical associations, and in the clinic as 'the organization of the hospital as a place of formation and transmission of knowledge [*savoir*]' (Foucault, 2007b, p. 151) – supporting the pervasive medical gaze that now needed to *isolate* the truth of disease in order to eradicate it completely. Whereas this gaze also incorporated the previous classificatory attitude, it also became a sort of 'speaking eye', for which what was visible coincided with what was recognisable and transmissible, in a (horizontal) coextensiveness of the truth of the visible and of the truth of the sayable, of 'speech and spectacle' (Foucault, 2003, p. 115).

The prominence of sight and the institutionalisation of control in identifiable places – first the hospital, then the laboratory – characterises what has been called 'surveillance medicine' (Armstrong cited in Nettleton, 2004, p. 663; see also Haraway, 1991, pp. 208 ff.), and is part of Foucault's analysis of disciplinary power – together, for example, with the panoptical gaze (Foucault, 1977; 2003; 2007b). Today, despite the persistence of previous forms of classification and control, a concomitant change in contemporary technospaces, where visuality has become a pervasive condition and the mobile-locative networks of information and communication increase both the possibilities of choices and exclusions (see Chapter 3), and in epistemological frameworks, where paradigms based on concepts such as complexity, emergence and hybridity have replaced organicist and mechanistic views of the individual and collective body, delineate a scenario in which information becomes a dominant metaphor.

The classical autopoietic body, the body of thermodynamics, was still organised around production and reproduction, accumulation and expenditure of energy. It was conceptualised as a body-as-organism, whose energy must be either channelled or expended (as labour force), if not maintained, so as to avoid the threat of entropy – an excessive accumulation of useless energy eventually causing the body's death. It was a body already seen as a field of forces but still entrapped in a closed shell, on/over which industrial capitalism built a whole 'technological, economic and biosocial order' (Parisi and Terranova, 2000). On the same premises, however, a different imagination of the body could be built. Entropy, rather than a negative tendency toward disorder and death, was progressively reconceived as the increased complexity of open systems, linked to their environments through feedback loops; a system could work via the incorporation of variations and disturbances towards the achievement of unpredictable states, as AL research would later show (see Hayles, 1999; Kember, 2003). In third wave cybernetics, emergence was substituted for the principles of homeostasis and autopoiesis of first and second wave cybernetics, respectively. Moreover, 'the post-cybernetic thought of complexity' could establish a 'positive correlation of information and entropy', going beyond the consideration of information as essentially anti-entropic (Ticineto Clough, 2003, p. 363).

Most importantly, as Luciana Parisi and Tiziana Terranova (2000) explain, the passage from the thermodynamic to the turbulent body in contemporary technospaces implies a rupture with the representational, in that the variable forces in which the organisation of the social opens onto a field of possibilities belong to the unexpected *performativity* of turbulent bodies rather than to their efficacious *definition*, either via equilibrium or via contradiction (Lazzarato, 2006, p. 177). Indeed, the consideration of information in terms of modulation rather than selection/closure of multiple informational flows requires an abandonment of the tridimensional space of Euclidean geometry – in which distance measures absolute differences – in favour of the qualities of 'dislocation, mutation and movement' of informational space as a space 'which does not presuppose the actions of a subject, but disassembles, reassembles and drags it out of itself' (Terranova, 2006, p. 52, author's translation; see also Guattari, 1995). Informational space defines not so much calculable space as the space that

> manifests an excess of sensorial data, a radical indeterminacy of our knowledge and a non-linear temporality which involves multiple variables and several interlaced levels of observation and interaction. Actually, space does not need computers to become informatic, although computers potentially make us more aware of the informatic environment as such. (Terranova, 2006, p. 52, author's translation)

As Deleuze, using a medical metaphor, writes in his 'Postscript of the Societies of Control' (1992), in the contemporary stadium of capitalist evolution, in which capitalism concerns the production of services more than of products and in which we participate in a technological passage from the energetic machines of disciplinary societies to the informatic machines of control societies, bodies live in a diffuse condition in which they are no longer moulded but modulated, caught in a continuous variation of '*metastable states*' (p. 5, emphasis added). This happens at all levels of society: factories are replaced with corporations, prisons with substitutive penalties, schools with perpetual formation, and hospitals with 'the new medicine "without doctor or patient" that singles out potential sick people and subjects at risk, which in no way attests to individuation – as they say – but substitutes for the individual or numerical body the code of a "dividual" material to be controlled' (p. 7).

As life becomes more and more entangled with information, noopolitics complicates biopolitics[9] through 'the widespread exercise and operation of information as a field of power relations that pervades all levels of society', transversally modulating the forces of collective memory, attention and affect (Munster, 2013, p. 133; see also Lazzarato, 2006). The management of information circulation, which is the possibility of handling life itself in its emergence as multiplicity – as both what resists (emergence) and is resisted (control) – becomes a central concern of networks, be they tactical or strategic (Thacker, 2008). Accordingly, Haraway (1991) considers the cyborg a 'text, machine, body, and metaphor – all theorized and engaged in practice in terms of communications' (p. 212). Its body is coded and organised as 'an engineered communications system' (p. 211) in which disease is frequently assimilated to 'a subspecies of information malfunction or communication pathology', as the contemporary importance of immunology and risk calculation makes clear (p. 212; see also Haraway, 1992; Franklin, Lury and Stacey, 2000; Nettleton, 2004; Thacker, 2008). Genetics and molecular biology reframe the body as carrier and a permeable transducer of information.[10]

Whereas governments increasingly employ networks of information and communication to develop national biodefense programs for disease surveillance (Thacker, 2008), these same networks also back up traditional places of medical knowledge so that the latter is potentially made more available and can be accessed and produced beyond traditional institutions. At the same time, the multiplication of screens and digital interfaces disseminates the image of the body and detaches it from actually located bodies: 'The medical body has not only *escaped* from the institutional discursive-material spaces of medical containment, 'it is also e-scaped in the sense that it is "viewable" through the electronic infoscape that is the internet', writes Sarah Netttleton (2004, p. 670). Notions of limit, horizon, inner and outer space are all redefined according to a new 'bodily environmentalism', as Sarah Franklin, Celia Lury and Jackie Stacey (2000, p. 40), drawing on Haraway (1992), show comparing, for instance, ultrasound images of foetuses, cosmic pictures of the planet earth and cell visualisations.

Overall, a fundamental ambiguity lies at the core of the material-semiotic apparatuses of contemporary technobiobodies, bodies which are characterised by their communicating boundaries rather than their internal integrity. On the one hand, they are subjected to new forms of more pervasive and mobile control,

---

9   The term biopolitics in Foucault's philosophy indicates the introduction of life and the living in history during the eighteenth century, where life is intended as a complex of forces that are considered productive, thus capitalisable, in themselves and which power forces, in this respect biopowers, aim at regulating and controlling (see Lazzarato, 2006).

10   Even though, as Nettleton notes (2004, p. 669), the first research on DNA still relied on a mechanistic model of the body rather than a systemic model.

which adopt the logic of networks and include colonising 'strategies of design, boundary constraints, rates of flows, system logics' (Haraway, 1997, pp. 211–12). On the other hand, the loss of integrity of multiply interfacing bodies encourages us to think of them as always localised in a 'system architecture' (p. 212) in which the context becomes 'a fundamental matter, not as surrounding "information", but as co-structure or co-text' (p. 214). The vital force of the body, autopoietic and relational at the same time, shows that living matter is already capable of information and that life is an unfinished, interactive process.

Thus, the management of circulation is (at least) a twofold, contradictory phenomenon which also brings two conflicting imaginaries with it. The image of the body disseminated and made available like never before, the promise of its total visual openness, is based on the representational assumptions of transparency and analogical correspondence. Here, bodies are never really mobilised: they are congealed in the representation of their diseases, working as reductive metonymies for the complexity of diseases which have been experienced and thus isolated and decontextualised. This representational imaginary, however, confronts a more performative one in which bodies are never fully readable or visible, never fully explicated, and in which different possibilities of displacement and interference emerge. An important consequence of this second aspect of the networked body is that the activation of its performative aspects takes place in what Latour (2005) has called a 'World Wide Lab' (p. 119), a World Wide Web of networks where the laboratory has been disseminated along with its traceable practices. Here, life and death are not contained in single bodies, but disseminated in an open circuit potentially accessible to multiple actors.

Issues of the circulation and control of information in spatial and bodily ecosystems are central to the work of Salvatore Iaconesi, an artist with a background in engineering, hacking and interaction design. They feature strongly, in particular, in the work of Art is Open Source (AOS), the duo that he founded together with his partner and wife, Oriana Persico, and which takes its name from a festival that Iaconesi organised inside the Linux Club in Rome in 2004. More than a duo, however, AOS operates as an activator of sociospatial networks via multimedia activities based on the employment of digital and ubiquitous technologies. Visualising existing networks and their relations, the flows, sites and actors of knowledge production and transmission, but also making data freely available and easily usable through a series of open access digital tools – the 'ubiquitous commons' – and workshop activities are the fundamentals of AOS's (2014a) practice. AOS (2014b) works according to a 'P2P ethnography',

> a participatory, performative approach, in which research and understanding require gaining awareness of one's position within the relational ecosystem (from cultural, emotional, aesthetic, perceptive, cognitive points of view) of the observed social group, and to establish or modify relations and interconnections both within the group, outside of it, and in-between, in fluid, dynamic, possibilistic ways.

Among the most interesting examples of this approach is the project *Human Ecosystems* (AOS, 2013–), a series of platforms for observing the movements, relations and emotions happening in cities in real time as they are performed in social networks, that AOS has been testing in several different cities all over the world.

In 2012, after two episodes of epileptic seizures, Iaconesi was diagnosed with a brain tumour. The small size of the edema and the smooth edges of the neoplasia – an index of a slow growth of cancer cells – visible from the images of his brain scans made with CT (Computerised Tomography) and MRI (Magnetic Resonance Imaging), showed that the tumour was a low grade glioma, localised in the language area of the brain. When, still in the hospital, Iaconesi asked for a printout copy of the scans to look at the image of his disease, he was denied access at first. A bureaucratic procedure had to be followed so that he could obtain the visible representation of his own brain. As a patient, he was experiencing a paradoxical condition: while, technically speaking, the scans in fact represented his brain, his images were also the place in which his body had been detached from his visible form, so that they seemed to exist independently from what they were conveying. Not only had Iaconesi's body all at once been flattened and transformed into its visual information, it had also disappeared as the originating place of such information, reduced only to a case, a number, a statistical occurrence.

Having experience as a hacker, Iaconesi felt the urgency to break the boundaries of the encoding surrounding his disease, to occupy his displaced body again and to bring his disease back into the complexity of his life's relations, desires, affects, and to look for a possible cure as well: 'I wanted the possibility for a society in which "to cure" doesn't mean an unidirectional thing (doctor to patient), or even a bidirectional one (doctor/patient collaboration), but a multidirectional, emergent, exploded, disseminated, non-linear one, engaging directly all of society. A peer-to-peer cure, an ecosystemic cure' (Iaconesi and Persico, 2015).

So, first of all, after leaving the hospital, he forwarded an official request to access his files. After a few days, he was given a CD which contained his digital medical records. However, when opening them, he realised that, although they were in a technically open format (in DICOM format, an acronym for Digital Imaging and Communications in Medicine), they were not suited for access and use by non-professionals as they not only required specific software and programming libraries to be readable but, once rendered readable, they also used such a specialised terminology that they remained practically incomprehensible. Thus, the first move of Iaconesi's process of 'reappropriation' of his brain was working to convert the files into easily readable formats, such as jpg., tiff., excel and html. In this way, they were ready to be shared and accessed by other people outside of the community of technicians and specialists in the medical field.

While breaking up a language code, Iaconesi was also starting to break a sociospatial one, bringing his disease out of the hospital walls – the antipsychiatric movement animated by the Italian psychiatrist Franco Basaglia in the 1970s being one of his sources of inspiration – but also out of the likewise regulated networks of medical information with their highly controlled and oriented flows. Starting to hack his disease, Iaconesi had in fact passed from a patient condition to an 'impatient' one (Bratton and Jeremijenko, 2008). Differently from cracking, which means breaking into an informatic system but usually with malicious intent, hacking is defined in *The Jargon File*[11] by its ethical, socially-oriented, attitude (Di Corinto and Tozzi, 2006, pp. 31 ff.). This also means that it 'concerns the way things are dealt with' rather than merely the kind of technologies implied in the process (p. 63, author's translation). To deal with things is primarily an ethical issue, as it has to do with engagement, care and accountability.

Thus, hacking a piece of information goes deeper than making its code legible to a wider number of people. It also extends to the process of foregrounding the otherwise illegible and invisible components concurring in the 'formation' of information. In fact, if ecosystems are also media systems, and if mediality regards both technologies *and* societies in their processes of co-emergence, then guaranteeing access to information alone is insufficient, given that when information moves, nothing stands still: the actors, the space in which information travels, the trajectory, the same piece of information, are all circularly mediated. Communicated information is information which has acquired a stable form, but since information is always also in-formation, the form of information is just the contingent closure of a more complex informational assemblage in which multiplicity – an intensive multiplicity of relations (see Chapter 1) – does not disappear, but is only constrained in its possibilities.

In this respect, as Iaconesi writes, a more inclusive definition of openness does not merely regard 'the lack of limitations', but

> the active presence of all of the conditions for the manifestation of freedoms and for the possibility to express oneself: the presence of an open format is not enough if it is not matched by its accessibility across languages, cultures and backgrounds, and if the importance of the desire to use it to express one's freedoms and liberties is not actively promoted, communicated and made perceivable. (Iaconesi and Persico, 2015, p. 61).

Hacking his disease's information, Iaconesi also abandons the patient's condition, one in which he is a readable-writable surface devoid of any agency, and becomes 'impatient', reclaiming agency but at a collective level. The artist Natalie Jeremijenko talks about the 'impatient' (in Bratton and Jeremijenko, 2008) when she describes her project *The Environmental Health Clinic* (Jeremijenko, n.d.), an actual clinic with a lab founded in 2003 and located in New York, in which people make an appointment to express their particular health concerns – never personal ones,

---

11   *The Jargon File* is a cooperative glossary of computer terms and netiquette handbook initiated by Raphael Finker at Stanford in 1975 and subsequently expanded over the years.

but regarding the environment in which they live – and are given 'prescriptions'. Clearly, these are prescriptions for actions at the local level rather than for drugs/ medicines. Even if they are personalised, different groups of people can work with the same piece of information and obtain different outcomes, performing their cures in a different manner. Moreover, the local performances they generate are reworked and 'reabsorbed' in the clinic circuit and can be used for further action.

In a way reminiscent of Jeremijenko's Clinic, AOS adopts the methodology of 'urban acupuncture' theorised by the architect and artist Marco Casagrande (2010), in its analyses and interventions in the urban environment (Iaconesi and Persico, 2014): this kind of spatial acupuncture starts from a consideration of cities as ecosystemic bodies whose mapping requires dynamic tools able to capture and design the movements of 'relational topographies' as they evolve, rather than to merely visualise them as they are. Combining time and space, this mapping methodology intervenes on the critical 'points of pressure' (Iaconesi and Persico, 2014) of the urban tissue in which changes can be generated, those links and nodes of the sociospatial assemblages where connections are made and can thus be unmade as well.

In *The Environmental Health Clinic*, politics is reframed in performative terms concerning the envisioning of alternatives rather than the picture of a present condition, because the artist not only works as a facilitator who triggers a process of sociospatial formation rather than as an expert who represents an already existing community; she also starts the ecosystemic health process 'from a field of pragmatic relations, rather than from a planned set of rules or fitness functions' (Parisi and Terranova, 2000). Like Jeremijenko, Iaconesi was not satisfied with making information available. He intended to make it performable, and in this manner *ethically* and *politically* usable.

Thus, after converting the medical files into a more readable format,[12] Iaconesi decided to upload them online and create the site *La cura. The cure* (AOS, 2012), whose subtitle reads *An Open Source Cure for Cancer*. Two concise headlines describe it. The first is 'A brain cancer. Some very personal Open Data. An opportunity'. The second reads, 'We can transform the meaning of the word "cure". We can transform the role of knowledge. We can be human'. At the same time, Iaconesi creates a three-minute video to explain his ideas with the help of his partner, who shoots it with an iPad: no script, no editing, just him talking to his imagined audience in front of the camera. In the video, Iaconesi explains how he will use the website and invites everyone to send all the possible cures they want, saying that he will publish all the replies, suggestions and proposals received as a feedback from the data sharing.

As soon as the video (to which other ones followed) was uploaded on YouTube[13] and on Iaconesi's Facebook page, as well as circulated in some mailing lists, it

---

12   A DICOM file viewer is now available on the website of *La cura* for those needing to open their medical data, or if they want to share them or even modify the software.

13   Further links and details can be found on the project website (AOS, 2012).

immediately went viral. To give only some numbers: the video had more than 15,000 views the first day, being also relaunched by major Italian newspapers and then international ones. The TED (Technology, Entertainment, Design) community, of which Iaconesi is a fellow, initiated global media coverage, relaunching the website and the video on the TEDblog. Various broadcast news channels uploaded Iaconesi's video to their websites, and CNN soon became the major referrer to *La cura*, that is, the initial website from which most of the visits to *La cura* website came from. To date, according to the draft version of *La cura. Working book* (Iaconesi and Persico, 2015) if we exclude more or less informal online and offline conversations and Skype calls, more than 900,000 messages have been exchanged, of which around one third were emails and 200,000 from social network sites, for the most part generated by the first YouTube video, and with approximately 200,000 people involved (most in single message conversations, a few in multiple message conversations). These findings are the result of a work of tracking and analysis that was conducted by AOS, monitoring conversational activities such as likes, shares, retweets, threads, hashtags and keyword recurrence, surrounding *La cura*. To do this, AOS used specific software and sociological approaches like Social Network Analysis, which also allowed Iaconesi and Persico to understand the sociospatial nodes and links, distribution and trends of the communities taking shape around *La cura*.

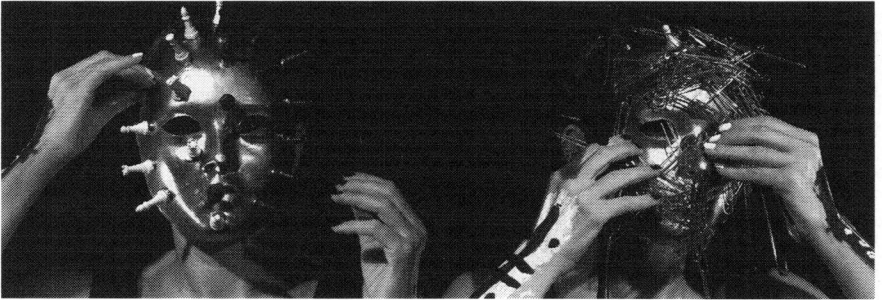

Figure 4.3   Photo of the performance Healing for La Cura (AOS), by Francesca Fini, 2014

Source: Courtesy of Francesca Fini.

The cures that Iaconesi has received and that have continued to arrive after his surgery in 2012 – more than 100,000 so far – vary from dance pieces and multimedia performances to poetry, from technical and 'alternative' medical opinions to reading tips, from 3-D sculptures of his glioma to religious or dietary advice. While these suggestions have not always been useful per se, they have nonetheless proved useful if taken together in their capacity to let *La cura*'s 'bomb of polysemic desire' explode (Iaconesi and Persico, 2015, p. 64). While all kinds of social actors, patients, artists, activists, professionals and hackers have been involved, one of the most difficult aspects of the project, according to Iaconesi, has been communicating the performative aspects of this work. Some viewers, particularly print media journalists, in fact tended to highlight the most sensationalistic and personal aspects of the cancer-art relation, whereas others mistook *La cura* for a new service for cancer patients. But AOS intends *La cura*, above all, as a performance, aimed at mobilising the interlinked networks of information, people and places in order to foreground what to cure and being cured means according to an ecosystemic approach. As Iaconesi writes, 'health is not something you subscribe to, it is something you desire and do, building it together with all the other members of your community. Health as a common, not a commodity' (Iaconesi and Persico, 2015, p. 74).

La cura is not just about knowing more or sharing in collective knowledge, it is about the practices of doing and undoing knowledge as well as the subjects of knowledge. As Haraway (1992, p. 319) explains, it is a question of cultural and technoscientific politics at the same time: it means creating new connections, new collectives, new encounters. What counts, but also who accounts for it, can only emerge in movement, through a distributed relational process in which different, diffractive articulations of the world take place (Haraway, 1997; Barad, 2007). Beyond all possible, more or less effective suggestions that flow into this collaborative work, what really cures in *La cura* is its dynamics of open, unending performance, which situates cancer on the side of life. As Iaconesi writes, 'this is one of the most interesting things about the performance: its ability to generate a world, some of which is still unknown to us, independent and autonomous in its beginning, life and development' (Iaconesi and Persico, 2015, p. 66). This is also why, in *La cura*, in-formation is considered as a qualitative, not quantitative issue, which in-forms reality according to unforeseen patterns and virtual twists. Here, life means leaving articulation 'accessible to action and interventions' (Haraway, 1992, p. 327), contrary to death as the closure of a system that retreats onto itself.

4.3 Significant Others

Differently from Iaconesi, who underwent successful surgery for his benign glioma at the beginning of 2012, the artist and activist Beatriz da Costa (1974–2012) dealt with cancer at various stages in her life from a young age, until a metastatic breast cancer which extended to her brain eventually caused her death in December 2012. Like Iaconesi, however, da Costa has left us with more than an artwork on the ecosystemic intricacies of contemporary technobiobodies, including narrating her experience after brain surgery in the work *Dying for the Other* (2012).

Both on her own and in collaboration, da Costa has worked with art groups such as the CAE and Preemptive Media, the latter of which she cofounded in 2003. While Preemptive Media especially experimented with mobile technologies, tracking data surveillance and monitoring environmental criticalities such as air pollution, her most renowned works with the CAE regard the field of what they termed 'contestational biology' (2002), for example, creating recombinant bacteria to question both the anxieties and the utopias surrounding the idea of natural purity, *GenTerra* (CAE and da Costa, 2001–2003), reverse engineering genetically modified seeds to expose their vulnerability, *Molecular Invasion* (CAE, da Costa and Pentecost, 2002–2004), or tracking the ability of genetically modified food to transgress national borders *Free Range Grains* (CAE, da Costa and and Shyu, 2003–2004).

Pigeon Blog (2006–2008), da Costa's most renowned solo work, used community-oriented data gathering devices that, through tiny sensors carried by flying pigeons, sent the retrieved information to an online server which mapped it to chart the level of air pollution in real time (da Costa, 2008). Data were made available to the public for further claims and actions, as in the custom of citizen science.[14] The work was inspired by a picture of a pigeon carrying a camera around his neck – a sort of proto-drone technology developed for aerial photography by the German engineer Julius Neubronner at the beginning of the 1900s and later used for military purposes during World War I – and intended to convert the use of mobile tools from top-down control to public utility. Not only did this work raise environmental concerns about the very polluted area of Southern California, it also foregrounded the importance of the interaction between human and non-human beings in urban space. An interest for other species and for interspecies dialogue animates many of da Costa's works, such as *A Memorial for the Still Living* (da Costa, 2009), an inquiry into species extinction which came with a related smartphone application – the *Endangered Species Finder* – and *Invisible Earthlings* (da Costa, 2008–2009) which, similarly to *GenTerra*, focussed on the social agency of the non-human, specifically microbes, and the way these invisibly perform our environment and can be handled by humans.

The answer that da Costa gave to the detractors of *Pigeon Blog* can be the starting place for understanding her approach. Actually, the piece was strongly contested by animal rights activists such as People for the Ethical Treatment of Animals (PETA), who not only accused da Costa's team of animal abuse, but also considered the operation not scientifically 'grounded' (da Costa, 2008, p. 380). The artist (da Costa, 2008) answered that the work was definitely not about animal rights, but about 'political cross-species art in action' (p. 380), a statement that will become more clear in the analysis that follows, and that signals how the common denominator of da Costa's activity has always been the search for 'relations of significant otherness' (Haraway, 2003, p. 8), more than the claim for absolute truths or rights. Moreover, on the same occasion, da Costa wondered why the

---

14 According to Iaconesi, this entails a definition of citizenship in which 'citizens are … more active, aware agents of their society, conscious that the well-being of their communities largely depend[s] on the extent and quality of the commons [*sic*] that they share, protect and work upon' (Iaconesi and Persico, 2015, p. 32). See also Chapter 3.

human-animal link should appear more critical when taking place in art rather than science, if not for a tacit epistemic privilege accorded to the second – and the belief in a gap between the two.

These questions were reframed and developed by da Costa in relation with the practices of cancer medicine in *Dying for the Other*. This is a video tryptich in the form of a three-screen installation shot in 2011 in the three months following da Costa's brain surgery. The video is part of the multimedia project *The Cost of Life* (2009–2012) which, as the title suggests, wonders about the value of life in between nature and culture, the individual and the environment. The project also includes: *The Life Garden*, the installation of a greenhouse containing plants, herbs and mushrooms which are known for their anti-cancer properties; *The Delicious Apothecary*, a medicine cabinet containing healthy herbs employed in the accompanying cooking class; and *The Anti-Cancer Survival Kit*, successfully crowdfunded on Rockethub but unfortunately only one year after da Costa's death.[15]

The simultaneous projection of *Dying for the Other* starts from the central of the three adjacent screens, where the letters of the central part of the title, 'for the', slowly start appearing in transparency, imitating a window decal, so that the viewer glimpses some moving figures behind the glass only the movements of which are distinguishable, not the shapes or the setting. After the words 'Dying' and 'Other' slowly materialise left and right, a fade-out to white leaves again the scene to the central screen, where in the place of 'for the' we now see a half close-up of da Costa, accompanied by a man beside her, who very slowly and unsteadily walks with an intent gaze along what looks like a hospital corridor, trying to keep an imaginary straight line in front of her.

The place where dying happens and the place of the other, that the words 'for the' relate in the video work, are not fixed ones. Sometimes we see da Costa in the place where dying happens, sometimes she is the other living in the place of the mice who die in the lab. A continuous interference is created by combining and alternating these images. For example, at the beginning of the video, while we look at the image of the artist walking along the hospital corridor, all at once, on the lateral screens, two slightly different angled images of the same mouse, kept in the hands of a lab scientist while injected, appear for a moment to disappear right after. Later, behind the façade of the American Cancer Society Hope Lodge Jerome

---

15 In the artist's mind, the latter should consist of a crate containing healing and 'feel good items' like teas and chocolates, a DIY garden kit and a related online database (*Anti-Cancer Survival Kit Plant Database*) providing references about the active principles and the growing instructions of the plants, considering their benefits for different kinds of cancer, and an illustrated artist book, plus several other online resources, such as an ebook, a database of scientific articles and educational games for mobile devices (Moore, 2013). A prototype of *The Anti-Cancer Survival Kit* was made for a solo show of da Costa's selected projects at the Laguna Art Museum in Laguna Beach, California, in 2013.

L. Greene Family Center in New York, whose door da Costa has just entered in the previous sequence, we meet the same researcher perfectly at ease in a laboratory, her movements quick and precise, handling a series of tools that resemble the objects of a domestic kitchen: a fridge, a stack of (petri) dishes, a dropper, several plastic and glass bins, and the mice, that appear again while taken from a cage to be injected, as well as while lying on a digital scale.

Without even realising it, as the cold light and colours remain more or less the same, the plastic case that we now see in the central frame is da Costa's pillbox, hers the hands that sort the pills into the little boxes assembled in a row that serves to organise the cure on a weekly basis. A kitchen, again, but da Costa's this time, with the artist organising her groceries and chopping kale, but also cutting her pills in half on a ceramic plate, beneath the highly disturbing image of a glass cage with what look like agonising mice inside. Her kitchen table does not appear very different from the lab table: perfectly tidy, it hosts only pill bottles, a pillbox, a knife and a white plate. While the researcher in the lab carefully weighs and sorts the mice, and with a detached voice explains that in general they use female mice as more fit ones to study breast cancer, da Costa continues separating and cutting her pills, as well as doing coordination exercises with her fingers.

Another disturbing interference takes place when we see the artist doing yoga and stretching on the floor virtually beside the researcher who first 'stretches', then dissects the mice bodies on the table to isolate their tumours and then extract them from their bellies. In the following scene, the researcher, having put the tumour on a petri dish, tries to keep it still with tweezers and to cut it open with a scalpel, at the same time that she, using a childish tone, underlines how 'teeny-tiny' the tumour looks – paradoxically, Haraway (FemTechNet, 2013) uses the same expression, but in a completely different meaning, noting that da Costa has learned to 'co-habit with the teeny-tinies' since her early projects. When the artist is shown while she exercises with her partner and tries to repeat sequences of numbers and words backwards, her shaky efforts with ordinary words contrast with the extreme self-confidence of the researcher who explains her procedure in the lab. However, when we compare the researcher's attempts to cut the tumour with da Costa's cutting her pills in half, this time the hands of the researcher seem hesitant, whereas da Costa's gestures appear firm and precise.

Just like the places of the other, the 'for', the place of articulation, is purposely left open in *Dying for the Other*. This is the place where both the activities of the observing subjects and of the observed objects converge, the place where 'crafting subject positions and ways of inhabiting such positions' are 'made … visible and open to critical intervention'(Haraway, 1997, p. 36). Here, viewers articulate their vision in relation with the other's, becoming aware of their visual practice.

Figure 4.4   Still from *Dying for the Other,* video installation by Beatriz da Costa, 2012

Source: Courtesy of Robert Nideffer.

Incidentally, hands are the most focalised body part in the tryptich. We see hands that hold, manipulate, cut, sort, squeeze, reach out, hands that do things, hands that kill and hands that cure, hands that kill to cure, hands that cure without killing, connecting hands and separating ones. In fact, hands shift the attention of the observer from the eye's contemplation to the hand's intervention. Through hands, emphasis is put on enacted practices.

Formally, the three screens projecting the synchronised video sequences reinforce the invitation to articulate vision and at the same time to be wary of such articulations. As magnifying glasses that work in reverse, so to speak, they amplify the edges of the frames rather than focus on their centre. Additionally, the rapid editing of the video sequences, which appear even more so when compared to da Costa's slow movements, does not permit viewers enough time to pause on one single screen, but switches their gaze from the one to the other: the representation of disease is distributed outside of the enclosed space of a single frame, which at the same time links the image of the artist's site and that of her disease to other diseases and other sites, establishing a series of visual connections that refer to material ones, and reinserting the personal dimension of disease in a broader frame.

In *Dying for the Other,* the preposition 'for' works as the conjunctive term that foregrounds the reversible relation of biopolitics and necropolitics (FemTechNet, 2013), of the dying and the living, of the partners that do not precede their relations but are the other of each other (Haraway, 2008b, p. 17; see also Barad, 2007). The mice are significant others (Haraway, 2008a) for da Costa, but at the same time, da Costa as cancer patient plays the same role for the mice, as does da Costa for other cancer patients that she stands for, and also for the art audience for whom she 'represents' her story. *Dying for the Other* shows the same 'relations of significant otherness' on which *Pigeon Blog* had already focussed: in fact da Costa's witnessing of her own story never disjoins from ethical attention for those 'companion species'(Haraway, 2003; 2008b) – be they bacteria, animals, human beings or technical tools – that share with the artist, and with us viewers, their stories of 'co-constitution, finitude, impurity, historicity and complexity' (Haraway, 2003, p. 16). The artist acknowledges such multiplicity as being already inside her body, a difference *within* that, diffracting the body through itself, poses the existence of exteriority in terms of relation rather than separation (see Haraway, 1997; Barad, 2012). It is not by chance that in many scenes da Costa wears a black cap and a very iconic black dress whose cuts, beyond resembling a skeleton chest case in a sort of *memento mori*, also evoke animal shapes, as if she were wearing a spider's legs or the striations of a butterfly's wings.

As Catherine Lorde (FemTechNet, 2013) notes in her conversation with Haraway on the occasion of the exhibition dedicated to da Costa at the Laguna Art Museum in 2013, more than 'who the other is', the question 'to what, exactly?' is paramount here. This latter question, apart from shifting the focus from identity to alterity, also reframes alterity as a material becoming, or better, as a *becoming with*. In a sense, *Dying for the Other* speaks about what it means to become with the other in practice, where practices, in this case, are the words, gestures, technologies, times and sites of dying.

According to Mol (2002), both the 'market model' of medicine, for which the patient is a customer that buys a cure, and the 'civic model', which considers medical interventions as juridical-like measures that the patient-citizen has the right to choose, rely on a 'politics of who' (p. 166) that presupposes a series of unquestioned elements: that the patient's will is already set and equal for everyone and must thus be preserved; that actions can be isolated from the histories that produced them; that data circulate between specialists and patients in a neutral, ever valid modality, as if they were exempt from contextual evaluation (p. 166 ff.). As we have seen, AOS's *La cura* already unhinges many of these premises, showing how information moves and what moves along with it when it openly circulates: it is not just a question of reaching more people, but of foregrounding the practices that make the transformation of sites, actors, and data possible inasmuch as they circulate.

Contrary to a 'politics of who', a 'politics of what' (p. 172) in medicine, says Mol, shifts attention from the choice of existing things to the process that makes them while making them worth choosing at the same time. Many differences intervene in enacting disease, depending on what acquires the status of good, what is good in what cases, what are the different effects of different treatments, and so on. The most important consequence of a 'politics of what' is that there is no order of things, but 'what to do' needs situating as well as inaugurating each time (p. 177). This is why, for Mol, doing, as doing disease and doing the knowledge of disease at the same time, becomes a matter of taking *sites* rather than *sides* (p. 178). Drawing on a concept of location mindful of feminist and ANT tenets, Mol specifies that 'the idea is not to celebrate localism instead of universalism. Instead, it is to keep track as persistently as possible of what it is that alters when matters, terms, and aims travel from one place to another' (p. viii).

As da Costa (2008) writes, the introduction of new media studies in the art field has not only redefined the boundaries of the latter, but it has also led to another role for the artist. The artists' toolkits have expanded, and thus their sites of action. The acquisition of skills in programming that, over time, has allowed many artists to acquire a political consciousness outside traditional institutions such as museums finds a parallel in handling the tools of the life sciences that have involved artistic practices in technoscientific experiments. The art field, enmeshed with other sites of knowledge production, has turned into an open and mobile lab in which to develop critical experiments of dissent.

In this framework, the answer given by da Costa to the critics of *Pigeon Blog* assumes a more complex sense: whereas, in fact, *Pigeon Blog* is not about animal rights – while not being abusive of them either – analogously *Dying for the Other* does not focus on patients' rights. Firstly, because it shows that rights do not exist in the abstract but are always constructed in a 'committed relationship' of mutuality (Haraway, 2003, p. 53); secondly, because it does not narrow the focus to the human side of the issue, nor does it create a human/animal contrast, but it compares different sites in which cancer is enacted using the creative process of art as a starting point to situate the analogous facticity of technoscience practices, and so diffracting the way they are usually represented.

Sites are locations where connections are either stabilised or mobilised in practices, but do not stand still. Thus, according to Mol (2002), the proper question should be: 'What becomes of objects when practices interfere with one another' (p. 121). This is why the artist also stands as the significant other of the lab scientist, and the artwork the significant other of the lab work. Actually, since the practice of art and the practice of science are 'natural sibling practices for engaging companion species', as Haraway states (2008b, p. 22), da Costa assumes a declaredly partial, situated position as a 'modest witness' (Haraway, 1997, 2000) that interferes with the apparently neutral scientist's position. Her seeing *attests* to the responsibility of taking a position in relation to the other, in contrast with the supposed objectivity of *testing* that the scientific lab conveys, which is questioned by the continuous interferences that this work creates.

At a visual level, in *Dying for the Other* da Costa accomplishes her cross-species art politics combining the two different manifestations proper of diffraction as the result of superimposing waves (Barad, 2007, p. 80): either she creates interferences between the scenes, alternating or paralleling her and the laboratory mice's images in context, as we have seen, and so connecting different sites where disease takes place, or she lets the narration proceed as in concentric circles that create a wave-like pattern: each subsequent sequence repeats and, at the same time, adds something to the previous one, not only going into further details but also widening the scenes' background, so as to give an effect of 'propagation' (Barad, 2007, p. 80) and amplification to the whole.

Exemplary are the sequences in which we see da Costa walking: initially, in the hospital corridor, we see her in a lateral close-up, the focus only on her face; then, the camera focusses on her back, and later, observing from a longer distance and from an almost frontal angle, we can also see da Costa giving her hand to the man beside her. Then, the set completely changes and the three screens all show different moments in which the artist, walking alone with the help of a walker and in the open, on a chaotic New York street, tries to enter a very heavy main door, an ordinary action which for da Costa, however, requires enormous effort. And again, the corridor appears, this time in a wider frame, revealing the hesitating steps of the artist in all their entirety, a scene which repeats twice, the second from behind, focusing on da Costa's head without her black cape so that we see her scar from the surgical wound vertically traversing her nape. Before the conclusion, the video again presents us with the lateral screens showing the initial image of da Costa walking along the corridor, but for the first time facing the viewer. Creating such parallel, expanding circles around the image of herself walking, da Costa actually goes wider *and* deeper into what this apparently ordinary action means for a cancer patient like her. She brings the personal to a higher level, diffracting her intimate account within a broader network of practices.

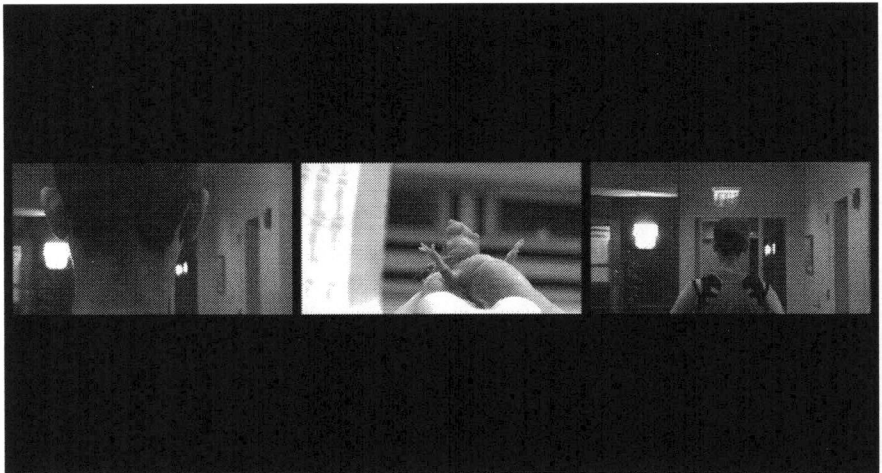

Figure 4.5   Still from *Dying for the Other,* video installation by Beatriz da Costa, 2012

Source: Courtesy of Robert Nideffer

This also translates into an entanglement of difference and repetition (FemTechNet, 2013), in which the search for balance is a key, challenging aspect. Balance characterises *Dying for the Other*, both at a technical and narrative level: the arranged comparisons and contrasts, the calculated alternation of visual angles, the shifts between subjects and places. Combined with interference and propagation, however, balance is never represented as an already conquered equilibrium, but as a dynamic tension which implies work, as we see from the very final scene of the video in which the camera smoothly moves along the artist's body standing in a room's corner, going from her feet up, while she is trying not to lose her balance, her eyes closed, her arms crossed on her chest, as in a meditation exercise. The ordinary stability of life is the performance of a series of successive and successful repetitions (see Butler, 1990). Specifically here, life seen through the lens of disease, and also diffracted through laboratory practices, loses its carefree naturalness and is shown to be built up throughout a constant maintenance work of natural-cultural concatenations.

Let us take, for example, the OncoMice™ that Haraway (1997; 2000) talks about, patented animals classified in databases and even sold in catalogues that, like other material-semiotic hybrids such as genomes, acquire a status of 'second-order objects', as Haraway calls them (1997, p. 99): beings that are made, although not made-up. In fact, their alienable bodies are the product of 'a mixture of labor and nature' (Haraway, 2000, p. 140), and for this reason, they need undergo periodic controls to see if they still fit the research aims, given that sometimes their genes can get lost in cell division and such mutations make them useless for experimentation. So, their identity is the outcome of a 'labor of maintenance' in which they are treated 'as if they were microprocessor chips' (p. 146). But da Costa's life as cancer patient appears to be the same: actually, 'mice and humans in technoscience share too many genes, too many work sites, too much history, too much of the future not to be locked in the familial embrace' (Haraway, 1997, p. 100).

What changes, however, is that the alienating processes through which OncoMice™ are made and subjected are instead diffracted by the artist through her embodied location in the story. This allows her to show how the maintenance of *all* life forms puts forth the necessary coimplication of multiple actors in the process. This is the sense in which Haraway (FemTechNet, 2013) talks of *Dying for the Other* as a work about 'assisted living', in which life can go on because it is distributed among several companions, including non-human ones, like animals (the lab mice, but also da Costa's dog that we don't see but which Haraway explains was trained to help her move around), plants (the healthy vegetables that da Costa buys and cooks), and objects (for example the walker that she needs to go outside).

Anatomy locates the topography of disease inside bodily boundaries, but a disease and thus a cure are part of life and, as AOS and da Costa's works show, are always done in practice, exceeding the realms of the seeable and the sayable within which bodies have been confined by traditional medicine. If the material-semiotic apparatuses of the latter have been employed for keeping the body – of the patient, of society, of knowledge – together, creating a subreptitious unity where, in fact, there were multiplicities, freeing multiple bodies from such constraints allows for 'coexistence side by side, mutual inclusion, inclusion in tension, *interference*' (Mol, 2002, p. 150, emphasis added).

In fact taking multiplicities apart, or rather making them 'implode' in a whole array of practices (Haraway, 1997, p. 68), reveals how practices as well as bodies do not organically or even mechanistically cohere, but rather assemble in open ecosystems. Like Guattari's (1995) biological and nonbiological machines, they are able to organise themselves; however, their autonomy is never, properly speaking, closed, since relations of alterity lead to a continuous 'disequilibrium', implying a 'radical ontological reconversion' (p. 37). Being, the being of such assemblages, is thus giving – that is, *being for alterity* rather than for the self. It is a generative process in which differences 'envelop each other' (p. 111).

In an ecosystemic approach like AOS and da Costa's, living and dying is not only for the other, but also with the other and with/in another. Thus, being cured and assisted also necessarily means assisting in turn, paying attention, taking care, being respectful. After all, as Haraway (2008b) underlines, this is the sense of looking at our companion species. Actually, species, seeing and respecting all share the same etymology:

> Looking back in this way takes us to seeing again, to respecere [*sic*],[16] to the act of respect. To hold in regard, to respond, to look back reciprocally, to notice, to pay attention, to have courteous regard for, to esteem: all of that is tied to polite greeting, to constituting the polis, where and when species meet. To knot companion and species together in encounter, in regard and respect, is to enter the world of becoming with, where who and what are is precisely what is at stake. (Haraway, 2008b, p. 19)

4.4   The Aesth-ethics of Diffraction

The systemic paradigm grows out of ideas of integration and equilibrium which the connective root of the word 'system' – *sys* – foregrounds and forgets at the same time, unravelling emerging differences in a seamless web, where technologies 'weave themselves' so deeply into the 'fabric of everyday life' that they eventually 'vanish' (Weiser, 1991, p. 19) in an invisible and ubiquitous background (see Timeto, 2013). Supporters of such 'fading' of technologies, drawing on Mark Weiser (1991) who coined the same definition of 'ubiquitous computing', consider this tendency a valid antidote to building virtual worlds inside computers because, as Weiser (p. 20) contends, cyberspaces end up being only maps, not territories. However, we have seen how the map/territory alternative is a false one in many respects, first of all because we never look at the world, be it physical or digital, and then at its representation, but we experience reality by means of representations that perform its enactments (see November, Camacho-Hübner and Latour, 2010).

---

16   The correct spellings of the Latin verbs are *specere* and *respicere*. For a further explanation of the etymology, see Zulato (2006).

In the name of a return to the real world, the invisibility perspective[17] in fact reinforces rather than weakens the assumptions upon which representational approaches rest, starting from the dichotomy real/virtual that it poses and ending with the presumption that the technological would only regard tools, rather than the very condition of the constitution of the social in technospaces.

Critical interventions regarding the integrated circuit (see Haraway, 1991; Guattari, 1995; 2000)[18] require exactly the opposite: that the machinism of technospaces, their joints, couplings and points of pressures, that is the '*withs*' where systems fix their thresholds, are not only brought to the fore but also brought towards instability, so as to prevent the 'entropic inertia' of control society (Guattari cited in Holmes, 2009; Deleuze, 1992), where everything that works properly must flow as smoothly and steadily as possible.

The dream of omnicomprehensive connectivity (see Graham, 2004a) overlooks the intra-active mediations (Haraway, 1997; Latour, 2005; Barad, 2007) out of which reality and representation co-emerge. On the contrary, our world is characterised by an 'immense machinic interconnectedness' (Guattari, 1995, p. 108) where being aware of the *schiz-*, the split at the core of all systems, as Brian Holmes (2009) writes commenting on Guattari's (2013) schizoanalytic cartographies, is as important as acknowledging their *sys-* component.[19] As a matter of fact, *com*panionships are not based on a levelling of differences in common identity, but on the enhancement of common differences in becoming, in which parts do not form wholes according to an additive quantifiable principle (Haraway, 2003, p. 25; see also 2008b, p. 83).

Haraway's (1992) political semiotics of articulation aims at retrieving the fissures and frictions through which realities assemble and that are found throughout technospaces, which are 'lost in doctrines of representation and scientific objectivity' (p. 313). Diffraction is Haraway's non-representational methodology for this political semiotics in which ontology, epistemology and ethics converge, as they will also do later in Barad's (2007) reprise. Compared with Barad's, however, Haraway always maintains a distinctive aesthetic stance, based on the persistence of vision delinked from geometrical optics.

---

17 For a very different, 'recombinant', version of the ubiquitous flowing of computing into the environment and its 'topological complication', see Thrift (2004).

18 Guattari (2000) distinguishes four semiotic regimes on which the IWC (an acronym of his for Integrated World Capitalism) is founded: economic, juridical, technoscientific and subjectification semiotics, which transversally cut the Marxian structure/infrastructure dichotomy since 'at present, IWC is all of a piece: productive-economic-subjective' (p. 48).

19 So, for example, somehow drawing on Guattari's ethico-aesthetic paradigm, Munster (2006) proposes to distinguish between connectivity and engagement when talking about the aesthetics of technospaces: engagement, in fact, also implies 'active confrontation' (p. 152) with the others, in terms of both active construction and responsible accountability, 'so that engagement with differences and others might be actualized in, rather than cede to, the political economy of connectivity' (p. 153).

This latter aspect rather pulls Haraway's approach closer to Guattari's (1995) theorisation of aesthetic machines, whose redefinition in fact creates a bridge between the aesthetic and the technological that is particularly relevant for re-articulating representations in technospaces. Distinguishing between the machinic and the mechanic, Guattari (1995) explains that machines work according to 'a double process – autopoietic-creative and ethical-ontological (the existence of a "material of choice") – which is utterly foreign to mechanism'(p. 108).

Firstly, let's see how machines are autopoietic-creative. The being of machines that Guattari talks about never precedes their actualisation: they are involved in a process of auto-modeling and meta-modeling – a modality of auto-spatialisation as well as of auto-representation that, while giving account of the concomitant production of signs and things, also escapes the overcodification of structures as well as the complete closure of the systems onto themselves (see Lury, Parisi and Terranova, 2012).[20] Even though, for Guattari (1995), 'the paradigms of techno-science place the emphasis on an objectal world of relations and functions' and privilege 'the finite, the delimited and coordinatable' (p. 100) of structures, which is sharpened in the 'hypercoordination' of the track-and-trace model prevailing in control society (Thrift, 2004, p. 185; see also Berardi Bifo, 2008, p. 33), technoscience's machinic Phyla[21] remain essentially creative.

Autopoietic machines do not belong to the order of objects: they are assemblages whose existence depends on their working, rather than on their essence, according to a 'creative practice and even an ontological pragmatics' (Guattari, 1995, p. 94). Creativity, for Guattari, is the eccentric force, the 'power of emergence' (p. 102) that aesthetic assemblages blatantly perform when they exceed the finitude of given forms by virtue of their 'affects and percepts' (pp. 100–101), but that *all* assemblages can perform when they work machinically rather than mechanically: that is, when they are performative rather than expressive machines. Here, Guattari considers affects as an ensemble of forces governing the human and non-human mediations of the technosocial (see Thrift, 2008; see also Grusin, 2010). In fact, in Guattari's words, 'affect is not a question of representation and discursivity, but of existence'[22] (p. 93). Affects, then, are the virtual forces conferring a processual and performative character onto realities and their representations.

---

20   For more detailed analysis, see Holmes (2009).

21   As Deleuze and Guattari (1987) explain, 'we may speak of a *machinic phylum*, or technological lineage, wherever we find *a constellation of singularities, prolongable by certain operations, which converge, and make the operations converge, upon one or several assignable traits of expression.* ... This operative and expressive flow is as much artificial as natural: it is like the unity of human beings and Nature. But at the same time, it is not realized in the here and now without dividing, differentiating. We will call an *assemblage* every constellation of singularities and traits deducted from the flow – selected, organized, stratified – in such a way as to converge (consistency) artificially and naturally' (p. 406).

22   As Massumi (2002) puts it, 'affects are *virtual synesthetic perspectives* anchored in (functionally limited by) the actually existing, particular things that embody them. The *autonomy* of affect is its participation in the virtual. *Its autonomy is its openness*' (p. 35).

In this scenario, aesthetic machines are considered by Guattari (1995) to be the 'most advanced modes' for the project of an 'ecology of the virtual' (p. 91). Interestingly, he explains that aesthetic machines cannot be apprehended through external categories or systems of reference (classical representation) but only through 'affective contamination': as 'limitless interfaces' (p. 92) that cannot be *represented* but only encountered, or, better, *taken* (pp. 92–3; see Chapter 1). Guattari's focussing on aesthetic machines does not imply an appeal to the social role of institutionalised arts, nor is he advocating an aestheticisation of society (pp. 102, 134). Rather, the aesthetic he talks about is the affirmation of a force of pure autopoiesis at its grade zero, so to speak – he calls it 'creationist nuclei of autopoietic consistency' (p. 105) – that while usually being more manifest in aesthetic machines, can be in fact retrieved in the heart of all machines. Thus, Guattari elaborates a 'proto-aesthetic paradigm' (p. 101) to approach this 'dimension of creation in a nascent state' (p. 102) that likewise traverses all domains and universes of value (p. 105). His primary intention is addressing creativity in order to recast the technoscientific domain beyond its actual realisations, on the side of becoming, intensity and virtuality.[23]

Assuming that an ontological capacity distinguishes both the autopoietic and the hetero-directed creativity of assemblages (Guattari, 1995, p. 107), aesthetic machines present a second aspect, which has to do with their 'ethico-political implications'. In fact, creation is not only a self-referential dynamic of machinic assemblages, but also means responsibility for the thing created, and that we take into account the fate of alterity in the process (p. 107; see also Haraway, 2003; 2008b; Barad, 2007; Lury, Parisi and Terranova, 2012). Conversely, being emerges from a process that from the very beginning presupposes relations of human and non-human alterity (pp. 100–101). In this respect, for example, Colebrook (2005) writes that art not only presents us with '"affectuality" – *or the fact that there is* affect' (p. 199), that is, with the thought of affect as the affirmative power that life has of desiring and 'differential imaging' (p. 199). Art also brings us to *experiencing* affectuality,[24] and this is art's ethical tendency, neither more nor less

---

23 The virtual, as Munster (2006) writes, does not belong to the order of representation because it does not derive from reality, nor does it belong to the order of simulation because it does not precede reality either: 'it is, rather, a set of potential movements produced by forces that differentially work through matter, resulting in the actualization of that matter under local conditions' (p. 90). Haraway (1992) reconnects the virtual to the original meaning of 'having virtue', that is, 'capacity', thus overcoming the dichotomy between real reality and virtual reality. According to Munster (2006), the virtual does not exclusively belong to new media technologies, but to bodies too; this means that the appeal to an embodied digital aesthetics is not an appeal to re-embody virtual technologies – which would still presuppose a duality, although undesired – but to conceive of 'new modes of techno-embodiment' (p. 115).

24 It would be interesting to confront Guattari's position with Niklas Luhmann's (2000) theory of art in the context of his system theory, in which the latter affirms the importance of putting aside a theory of causality as well as of mere interpretation when

than the 'pleasure of conjunction' at the core of any composition (Berardi Bifo, 2008, p. 35).

Let us for a moment back up a few steps, reconsidering Haraway's articulation of representation already examined in Chapter 2. Whereas the conflation of representation and spatialisation has usually been employed to silence the dynamism of both, in Haraway's theory representation is articulated because it partakes in a temporality which intimately enfolds in space as well as unfolds with space. The relation that Haraway (1991) poses between locations and figurations is not a static, dichotomous one, but one based on dynamism which implies time in many respects: if 'positioning is ... the key practice grounding knowledge organized around the imagery of vision' (p. 193), at the same time 'figures' are always tropic' as they 'involve at least some kind of displacement' (Haraway, 1997, p. 11).

Within the hybrid 'nature' of nature, *topos*, the commonplace of our discourses or the meeting point of public culture, and *trópos*, etymologically 'what turns', that is 'figure, construction, artifact, movement, displacement', cannot be disjoined (Haraway, 2008a, p. 159; see also Haraway, 1992). *Topoi* are unfinished, partial locations, whereas *trópoi* are embodied and non-transparent figurations. Just as *topoi* move, so do *trópoi* get settled and inhabited. Equally, *topoi* do not stand on the material side more than *trópoi* stand on the semiotic one: the processes through which they emerge is a process of differentiation, in which they reciprocally translate the one into the other. Location is oriented towards its turning edges (see Haraway, 1991, p. 195), while *trópoi* look for situatedness. As for space, time is never 'just there' (Haraway, 1997, p. 41), but is entangled with both space and representation as differing-deferring force (see Doel, 2010). Displacement differentiates location, like diffraction differentiates representation, so that both are distanced from the grip of identity and self-evidence.

In such performative spatio-temporal practices, vision plays a key role for Haraway, provided that it is delinked from the metaphysics of light and enlightenment to which it has traditionally been associated. Vision intra-acts (Barad, 2007) with reality *as* this *comes to* matter, which means that just like reality, even when representations stand still, they always maintain a diffractive capacity,[25] a nucleus for differentiation like Guattari's (1995) machines. Each *turn* in space-time brings with it a potential for change and invention that representations conjointly and dynamically *con*figure (see Suchman, 2012). Multidimensional cartographies are also multi-temporal ones (Haraway, 1991; Braidotti, 2003). In fact, revision regards the past but also the location towards which one re-turns in order to turn it over;

---

considering art objects, and instead adopting a second order perspective in which 'the author of an artwork adapts to the beholder in the same way as an observer anticipates another observer, and ... the artwork ... not only mediate[s] between diverging observational modes as they arise but also needs to *generate* such diverging perspectives to begin with' (p. 76, emphasis added).

25   See infra, note 23.

observing means engaging in the mattering of actuality; situated perspectives offer objective vision while also assuming a visionary 'orientation' towards envisioning the future. Additionally, for Haraway, vision is also always non-unitary for at least three reasons: firstly, it is embodied and so is stereoscopic (contrary to the abstracting tradition of linear perspective that relied on one eye only); secondly, it is situated and so it is partial; thirdly, it combines the objective and the visionary, the rational and the imaginary, and so it is cross-eyed (Haraway 1991).

Given these elements, a comparison between Guattari's and Haraway's aesthetics becomes possible. In Haraway's thought, the creativity of imagination co-exists with and, at the same time, exceeds the objective in both seeing and knowing. At issue is not a simple appeal to bridging separate domains of knowledge production according to a logic of transitivity and equivalence (see Mol, 2002; Code, 2006; Barad, 2012), but the very condition that subtends all world-making practices, their 'power of emergence' (Guattari, 1995, p. 102) that always contains the possibility of 'a twist in the ... kaleidoscope' of reality, to recall the diffractive metaphor that Code (2006, p. 121) employs when she writes about the creative negotiations that the '*art* of medicine' puts into play.

In Haraway's thought, creativity makes knowledge and imagination, epistemology and aesthetics, companion species for tracing common genealogies and intervening in making the present as well as imagining the future. There is no need to keep them apart, once the process through which certain knowledges emerge has been retraced and reactivated in its contextual performance as itself the enactment of an imaginary. Indeed, the creativity of imagination, which has nothing to do with fantasy but has much to do with facticity, re-situates knowledge inside an ecological network of practices which include places, relations, affects, bodies. In ecological terms, creativity is the 'process-oriented' aspect of reality-making practices at all levels (see Barad, 2007; Lury, Parisi and Terranova, 2012; Hughes and Lury, 2013), the possibility of envisioning 'a praxic opening-out' in actuality (Guattari, 2000, p. 53), 'how things change rather than how things are' (Hughes and Lury, 2013, p. 792). Guattari (2000), for example, uses the diffractive French verb 'dériver' to indicate how creativity intervenes in the formation of ecosystems which are never given in-themselves but are actualised (p. 53; see also Berardi Bifo, 2008). In fact, using the example of artistic practice, he writes:

> This new ecosophical logic – and I want to emphasize this point – resembles the manner in which an artist may be led to alter his work after the intrusion of some accidental detail, an event-incident that suddenly makes his initial project bifurcate, making it drift [*dériver*] far from its previous path, however certain it had once appeared to be. (p. 52)

If the relation between imaginaries and knowledges, as Haraway and Guattari make clear, is one based not on expression but on performativity, imagination must always be situated so as to mobilise creativity in practices and, at the same time, remain distinct from actualised imaginaries. In fact, just as no absolute knowledge exists, so no absolute imaginary exists as its counterpart. For this reason, Marcel Stoetzler and Nira Yuval-Davis (2002), like Guattari (1995), link the performative force of imagination to embodied affectivity, drawing on Spinoza's concept of affect (1989; see also Braidotti, 2002a; 2003; 2006; Deleuze and Guattari, 1987), but more specifically on Haraway's (1991) notion of a split, non-identical self: here, rationality and imagination are not separate faculties but '*dialogical moments* in a multidimensional mental process' (Stoetzler and Yuval-Davis, 2002, p. 324).

For Stoetzler and Yuval-Davis (2002), situated imagination is both a complementary condition and a product of the process of knowledge construction, one which '*stretches* and *transcends*' (p. 316) standpoints in futurity and makes them available for commonly constructed social and political goals. In particular, like Code (2006), they also draw on Cornelius Castoriadis's (1994) notion of creative imagination, for whom imagination retains a functional aspect for the social to which it guarantees a margin of freedom that cannot be predetermined. This helps them redefine imagination as both an epistemological and social category that extends beyond actuality. Their point is that an *in se* of imagination does not exist, but that imagination is a differentiating force both without and within existing imaginaries. As they put it, 'whatever we consider to be a currently impossible, but perhaps desirable, goal or value is always modelled – *ex negativo* – on whatever we perceive and imagine to be the actual and the possible in existing society', however different, contradictory and uneven such perceptions and imaginations can be (p. 321). Actually, this depends on the way assemblages assemble, whether as mechanisms or as machines, that is, whether they catch their aesth-ethic capacity in the 'sys' or liberate it in the 'schiz' of their autopoiesis.

An 'aesthetic processual paradigm' (Guattari 1995, p. 106; 2000) focusses on the creativity of imagination when it works at liberating the connections of technospaces, where the flows and forces of contemporary interconnectedness have also been valorised as models for the new economy of the integrated circuit (see also Deleuze, 1992; Berardi Bifo, 2008, p. 33). Drawing on a position which is quite popular among new media theoreticians, especially from McLuhan (1964), Guattari (1995) sees in interactivity, provided that the organisation of society changes accordingly, a decisive step towards a return to orality which will eventually reinstall another kind of dialogue with machines based on human-machine contact and affective conjunction rather than codification (p. 97). However, the issue is not so much the choice of the oral over the written, or affect over code. In fact, such terms are not essentially dichotomous. Rather, since new technological mechanisms alone do not bring about a 'refoundation of political praxis' (Guattari, 1995, p. 120), new machinic assemblages, Haraway's *different mediations*, are necessary too. Hence, a different aesth-ethics of technospaces needs to take into account the differential relations among its components, whose modalities of being always depend on partial connections and relations of alterity.

Accordingly, explaining Castoriadis's notion of 'instituting imaginary' as the same possibility of critiquing an 'institutive imaginary' – that is, an imaginary which has acquired stable meaning and defined the shape of common experience

– from within, rather than as an oppositional imaginary from without, Code (2006) warns us not to adopt a dicothomic approach not only when talking about imagination versus knowledge, but also when talking about a hegemonic – institutive – imaginary versus an alternative – instituting – imaginary (p. 33). In fact, 'all thinking, knowing, doing occurred *within* an imaginary' (p. 206), including imagining itself. Just as knowledge-making practices exist only in specific contexts, their being abstracted and decontextualised happens because of the imaginary that has emerged and endured as the dominant one in a precise spatio-temporal situation.

Some imaginaries are grounded, some overturn grounded ones. So, for example, Haraway's articulation of *topoi* and *trópoi* is also further elaborated by Yuval-Davis (2006) as the double movement of 'rooting' – for locating – and 'shifting' – for dislocating – that has characterised the shift from identity politics to transversal politics in feminism in recent decades. Once again, it would be a mistake to think that imagination is only involved in the act of shifting. It is, in fact, equally 'rooted': like knowledge and imagination, so rooting and shifting are already co-implicated. In this respect, contrary to what Barad (2007, p. 381) contends, it is very difficult to assert that a substantial difference exists between Haraway's statements that optics concerns the politics of positioning (1991) and that diffraction is about making differences in the world (1997): clearly, both rely on a continuous search for common grounds and for their displacement before commonality completely closes onto itself, excluding relations of alterity (see also Berardi Bifo, 2008, pp. 104–5).

As Haraway (1991) notably puts it, the 'task of reconstructing the boundaries of daily life' starts from an acceptance of the technologies of our lives, which put us in 'connection with others [and] in communication with all of our parts' (p. 181). The awareness of the heterogeneous configuration of our realities is also the awareness of their ongoing heterogenesis (see Barad, 2007) both coming from the past, as refigurations, and tending towards the future, as prefigurations. So, configuration is a self-reflexive tool for drawing boundaries and a generative one for 'recovering the heterogeneous relations that technologies fold together' at the same time (Suchman, 2012, p. 48).[26] As Suchman (2012) sharply recapitulates in Harawaian terms, 'an orientation to configuration reminds us to reanimate the figures that populate our sociomaterial imaginaries and practices, to examine the relations that they hold in place and the labours that sustain them, and to articulate the material semiotic reconfigurations required for their transformation' (p. 55).

Haraway's (1992; 1997) aesth-ethics of diffraction is 'a critical practice for making a difference in the world' (Barad, 2007, p. 90) that invests multiple spatio-temporal dimensions as it not only draws on a respectful recognition of a common evolution with the other of our reality; it also envisions the possibility of 'joint futures' in shared technospaces (Haraway, 2008a, p. 7). The entanglements of

---

26 Suchman (2012), drawing on Law, talks about configuration as a 'method assemblage' (p. 55).

diffraction make visible the responsiveness as well as the responsibility of those involved in the world's materialisations: here, representing is very far from the activity of a subject that contemplates and masters the object from a safe and invisible distance, and is rather a joint practice based on mutual relationships. In sum, on an ethico-political plane, keeping situated knowledge and situated imagination together does not so much serve to find new representations – at least not in a representational sense – as to perform new practices according to an alternative imaginary (Haraway, 2008a, p. 177). It means embracing the aesth-ethic forces of our lives intended as 'the way bodies perceive each other in the social field' (Berardi Bifo, 2008, p. 32), to be able 'to be other, to become other, to take pleasure in the other' (p. 165). Imagination, as the virtual capacity of autopoietic systems to *turn* commonplaces towards unexpected openings, is the starting point for a responsible praxis based on a commitment and attentiveness to the other, both within and without.

Opening Conclusions
Performing Represent-Actions

As we have seen, a traditional representational approach, both in the arts and sciences, relies on a series of binaries that keep the subject and object of representation separate as well as unquestioned, assuming that their respective spaces already exist and that representation is an instrument with a constitutive and pre-assigned function that the subject can pick up, as if it were from a toolbox, whenever the need arises. In this case, representation would work at bridging, so to speak, the voids of the subject/object gap, creating a series of analogical correspondences that go back and forth between the two opposed realms of signs and things, as a procedure that variably reflects or transcends the distance that it, nonetheless, contributes to keeping open. However, it has been precisely this idea of space as lying *between* the subject and the object as a separating plane, rather than as passing *through* them as a relating field, that has strengthened the toxic equation of representation and spatialisation.

In this book, several crucial binaries that have supported the representational paradigm have been taken into account, starting from the opposition between space and place, in which the latter has been overloaded/overburdened with the experience of authenticity, and the former has been quantified and immobilised while emptying, with a conjoined move, both space and its representations of their temporal dynamism. This has largely foreclosed the possibility of considering the interrelation between representational practices and the practices of space, which are possible only when navigational, 'auto-modelling' cartographies are employed that are not exterior to the space they represent but which in-form it as soon as it emerges out of its contextual arrangements.

The constitution of a comprehensive visual field, based on the complicity between traditional representational practices and the creation of the spaces that they reflected, has had many theoretical and ideological consequences that can be summed up in the paradoxical belief in the adequacy of representation, on the one hand, and its correspondence to an 'out-there' reality sanctioned with this same act of faith, on the other. Having imposed the representational imaginary of Western Modernity as the dominant paradigm, this not-so-ingenuous play of cross-references has also continually concealed its limits. Or, we could also say, it has worked *thanks to* this concealment, disguising the productive, relational quality of spatialisation activities and suppressing any consideration of differences both inside and outside, while also perpetuating itself along a chain of rigid dichotomies.

However, the more productive answer to face our inability to represent today's spaces is not to stop using vision together with representations or to regret a lost set of signs, longing for their return – not least because we live in a visually saturated media environment where vision has become pervasive and surrounding. An articulation of vision and representations as highly mediatised practices certainly seems more appropriate for performing the articulations of contemporary technospaces, being, and making them in turn, contingently practicable as well as imaginable. In technospaces, the social and the technical, humans and machines, enmesh in reciprocal mediations and assemble in only partially connected formations; thus, they are always open to change. In this respect, technospaces are a privileged field from which to observe the effectivity of performative representations that envisage and rematerialise the *creative capacity* of sociotechnical assemblages.

The return to the real of lived, though hypermediated, experience – as opposed to the separate dimension of digital virtuality, which the proliferation of ubiquitous interfaces and new ICTs like locative mobile devices often convey – if seen from another angle, also appears as the culminating point of a process of dematerialising information through a concatenation of frames, going from the wall to the window to the screen. In fact, contrary to what may intuitively appear, a cumulative although unending 'enframing' activity follows the increasing mobility of the flows of pervasive computation and models itself on the same undulations of information networks in order to canalise them according to paths of valorisable and even predictable signification.

So, mediations of technospaces happen to be simultaneously multiplied and erased along a series of arrangements and tracks that, once again, separate mediations, as reality-making performative operations, from media devices as objectual vehicles of reality: this not only again proposes the binary of the social and the technical, but also reduces media to an issue of representation and representability, as mechanisms that can be detached from their productive transactions from an outside that they would not include. However, it is precisely this necessity of an incessant locking of mediations which reveals that a universal, unfinished interconnectedness exists beyond and between fixed edges.

As the theorists and artists that I have discussed in this book agree, ICTs are crucial for practising and imagining our lives in the circuit of technobiopowers that innervate human and non-human bodies in technospaces, from the molecular to the molar. They can be used as encoding tools to quantify, signify and control the heterogeneity of informational flows, measuring both their current and probable directions. At the same time, these same flows demonstrate that the routes of information cannot be so easily abstracted by the embodied and mediated realities with which they intra-act, whose opacities and proximities interfere with the dream of representational clarity, bringing to the fore the dynamic entanglement of signs and things.

Therefore, this fluidity and continuous reworking of the boundaries of contemporary technospaces entails representational practices that no longer rely on the instruments of classical representationalism. Actually, if by using a traditional representational framework the boundaries of technospaces are disguised while a series of dichotomies are maintained or reinforced, observing and putting these boundaries in constructive tension means abandoning the binary logic that keeps exteriority and interiority separate, while contradictorily occulting the seams and junctures that reveal them as co-implicated and make them differently separable each time.

For all these reasons, I have chosen to focus on the theoretical and practical consequences of adopting a performative, non-representational approach to both space and representation, as a way of reworking representation so as to interfere with, and possibly overturn, the mechanisms of traditional reflexivity and simultaneously intervene at the level of reality-making. For the purpose of my argument, I have drawn on the legacies of feminist politics of location revised through Haraway's thought, and particularly her semiotics of diffraction and Barad's interpretation of it, but also on the cues arising from Foucault's spatial theory, ANT and non-representational methodologies, and not least Guattari's ethico-aesthetic paradigm. In fact, the relational rather than analogical linkage between spaces and representations, variably foregrounded by such approaches, pays attention to the ecosystemic interconnections of contemporary technospaces whose description and practice abandon any transcendental pretension of representational vocabulary and demand a relocation of representation.

Diffraction is both the non-representational, although optical, figuration and the methodology proposed by Haraway for articulating situated knowledge together with situated imagination, and their unceasing alternation of location and displacement. Classically intended, diffractions manifest the behaviour of waves that, differently from the way particles each have a precise space (position) and time (momentum), interfere and overlap when impinging on the edges or passing through the gaps of an object that detects them. But if we, for example, consider the waves produced by light, the approaches of geometrical optics and physical optics to the issue are very different. The former has only considered what happens when light encounters an obstacle, as said above, without questioning the nature of light regarded as a tool. The latter, on the contrary, has interrogated the dual, entangled nature of light as wave and particle through a series of experiments, in which diffraction has served to expose where the effects of entanglements appear rather than enlightening already existing, possibly entangled, objects. And, significantly, it has also uncovered the entangled function of material-semiotic apparatuses of observation.

So, if traditional representationalism firmly believed in reflexivity and, consequently, worked by separation and opposition of the observer and the observed and viewed observation as an instrument of the former, a diffractive methodology, which is also a different theory of mediations, on the contrary follows the co-implications of the observer and the observed, making of observation an intra-active relation between them which, like the related phenomena, does not pre-exist the phenomenon of relation itself.

Since we configure our world and establish connections with it through our ways of seeing and knowing, diffraction, so intended, does not simply regard our visual field but is a practice that invests our being in the world, our imaginary and our actions at the same time. The productive interference caused by diffraction discloses the unforeseen possibilities of an incompleteness that does not manifest a diminution of reality but, on the contrary, its intrinsic reserve of creativity, if by this term we intend the always differently actualisable capacity that contingent limits can alternately contain or release.

Diffracting representations always starts from locating them, abandoning the safe distance that kept the knower's gaze invisible and thus impossible to look back to: location, especially if we follow the meaning that the notion has acquired in the theories discussed here, starting from feminism onward, goes beyond a merely geometric or territorial definition, encompassing a complex idea of positionality whose implications are of the onto-epistemological as well as the ethico-political order, and whose outcomes are, as a matter of fact, both imaging and imaginative. Locations are material-semiotic dimensions, recognised as produced and productive, which dismantle any scalar logic and its hierarchies, stretching beyond their boundaries which they cut through and mobilise, but never erase.

As in the case of representation going back to location does not mean dwelling in a safer and handier dimension from which to reformulate a new spatial language for confronting the collapse of traditional spatio-temporal coordinates, nor yearning for or rediscovering a dimension of authenticity in which to eventually reside after losing orientation. On the contrary, the very moment that location is gone back to, its destabilising process starts. Actually, grounding and displacing are not two separate movements, they are interdependent. Locations are spaces of engagement in that they can never be taken for granted, just like their representations. Locations are figured and can be refigured in the same moment that they are practised and transformed by means of alternative encounters that turn their 'commonplace'. In fact, diffracting representations transform locations, disobeying the logic of the Same and engaging with differences in contingent articulations.

Indeed, when figurations of space are not delinked from the processes of spatialisation, as when information is not separated from 'mattering' matter, representations can be grounded in the lived spatio-temporal realities with which they engage and whose boundaries they also perform in mutable configurations. This brings to the fore the generative forces that realign the practice of representing spaces and the practice of situating representations inside a topology of variations, in which continuous *represent-actions* take place. Accordingly, depending on how they are employed – whether to keep existing configurations together or to disassemble them for further reconfiguration – performing representations can contribute to either locating or displacing actual places, either model or dissolve defined forms, either building or losing existing connections inside a common field of infinite and un-predetermined actualisable possibilities.

A simultaneously displacing and diffracting move is required, so that the space of one's own situatedness and the representation of one's own space leaves room for the other that is already within, but is impossible to either perceive or figure from the Subject position. The openness to alterity and heterogeneity that the proposed performative relation of space and representation positively confounds also allows for the adoption of a recombinant perspective in which the mediations inside and among human and non-human beings in technospaces leave room for the creative potential of un-predetermined joints, functions and actions.

In an ecosystemic framework, in which systems are characterised by relational openness and intensive multiplicity rather than self-sufficiency, identities are forced to an impasse of perpetual conjunction at the interface. Here, representational practices find themselves entangled in local contaminations, haunted by the pleasure of connection as well as by the always possible danger of disconnection. Such mediating zones of interaction and counteraction, where mirrors lose their reference points precisely when the gaze paradoxically looks back at them, are the *constraints*, or the *cuts*, where vision recognises its embodied partiality and envisions its inseparability from alterity. In front of these paradoxical heterotopias, self-reflexivity goes beyond the awareness of a symmetry between the subject and the object of knowledge. As an activity of creative ecosystems whose autopoiesis never completely closes onto itself, being rather a matter of co-emergence at the boundaries, self-reflexivity becomes the contrary of self-enclosure, implying the responsibility of opening to the other that sets the foundations for an ethico-political understanding of the *active implication* of the subject in the object's world.

The epistemological and aesthetic dismissal of the paradigm of distance, which a traditional representational approach has relied on both in the arts and in the sciences, gives way to modes and spaces of articulations of meaning and being undecidable in advance, but that can be engaged precisely because of the conjunctive, mediated character of technospaces. Given that connectivity is an abused word when talking about technospaces, understanding what it means and how it works in these hybridly articulated contexts becomes paramount. This understanding goes hand in hand with the awareness of the mediated and mediating processes of our information and communication environment. If we consider technologies not as mere instruments that influence social changes or that are available for social uses, but rather in terms of a co-existence of humans and machines, then we can devise a different social politics that replaces hierarchical relations of mastery with reciprocal relations of care and *respect*.

In contemporary technospaces, mediation is distributed everywhere in the interpenetration of the social and the technical, the actual and the figural, the material and the informational, across unending interfaces that actively perform and transform such relations. But mediation is precisely what disappears when we adopt a classical representational framework, where everything stands on one side or the other, but never in the middle.

Instead, in a context of shared agency and diffused relationality between heterogeneous beings, connectivity becomes an ethical and political issue because it always requires an ability to actively engage with differences in respectful companionship. Machinic assemblages live inasmuch as they work. Their being is their doing. Just like there are no external referents that validate their existence from the outside, so there are no representations that can fix their meanings or forms from a distance, once and for all. In technospaces, their processual creativity can be either canalised and codified, to control the modulation of contemporary flows, or let free to traverse multiple domains, so as to intensify autopoietic nodes.

Here is where the aesthetic encounters the ethical: so intended, creation means responsibility for the things created. It is not an isolated act of the individual, but a gesture of mutuality that takes into account alterity at the very moment of its appearance as an inassimilable, though engageable, in/appropriatedness. For these reasons, analysing the theoretical and practical implications of reconceiving the link between representation and spatialisation, with a particular focus on the representation of technospaces, I have offered the reader my situated, declaredly partial account, and also my respectful declaration of engagement with the other partialities with which I connect everyday.

Bibliography

The Act of Killing, 2012. [Film] Directed by Joshua Oppenheimer. USA: Final Cut for Real.

Adams, P.C., 1997. Cyberspace and virtual places. *Geographical Review*, 87(2), 155–71.

Agnew, J., 2005. Space: Place. In: P. Cloke and R. Johnston, eds. *Spaces of Geographical Thought: Deconstructing Human Geography's Binaries*. London: Sage, pp. 81–96.

Alaimo, S. and Hekman, S., 2008. Introduction: Emerging models of materiality in feminist theory. In: S. Alaimo and S. Hekman, eds. *Material Feminisms*. Bloomington, IN: Indiana University Press, pp. 1–19.

Amin, A. and Thrift, N., 2002. *Cities: Reimagining the Urban*. New York: Wiley.

Anderson, B., 2014. *Encountering Affect: Capacities, Apparatuses, Conditions*. Farnham: Ashgate.

Anderson, B. and Harrison, P., 2010. The promise of non-representational theories. In: B. Anderson and P. Harrison, eds. *Taking Place: Non-representational Theories and Geography*. Farnham: Ashgate, pp. 1–34.

Anthias, F., 2002. Beyond feminism and multiculturalism: Locating difference and the politics of location. *Women's Studies International Forum*, 25(3), 275–86.

Anzaldúa, G., 1987. *Borderlands: The New Mestiza = La frontera*. San Francisco: Spinsters/Aunt Lute.

AOS, 2012. *La cura. The cure*. [website] Available at: http://opensourcecurefor cancer.com [Accessed 24 February 2015].

AOS, 2013–. *Human Ecosystems.* [multimedia project and website] Available at: http://human-ecosystems.com/home [Accessed 24 February 2015].

AOS, 2014a. *The Fundamental Posts on Art is Open Source*. [online] Available at: http://www.artisopensource.net/category/fundamentals [Accessed 24 February 2015].

AOS, 2014b. *Communication, Knowledge and Information in the Human Ecosystem: p2p Ethnography*. [online] Available at: http://www.artisopensource. net/network/artisopensource/2014/07/30/communication-knowledge-and-information-in-the-human-ecosystem-p2p-ethnography/#comments [Accessed 23 February 2015].

Appadurai, A., 1996. *Modernity at Large: Cultural Simensions of Globalization*. Minneapolis, MN: University of Minnesota Press.

Aristotle, 2012. *Rhetoric*. Translated from Greek by W. Rhys Roberts. [ebook] Available at: http://ebooks.adelaide.edu.au/a/aristotle/a8rh [Accessed 5 April 2015].

Augé, M., 1995 [1992]. *Non-places: Introduction to an Anthropology of Supermodernity.* Translated from French by J. Howe. London: Verso.

b.a.n.g. lab and EDT 2.0, 2013–. *Transborder Immigrant Tool, a Mexico/US Border Disturbance Art Project* [blog] n.d. Available at: http://bang.transreal. org/transborder-immigrant-tool [Accessed 23 February 2015].

Barad, K., 1999. Agential realism. Feminist interventions in understanding scientific practices. In: M. Biagioli, ed. *The Science Studies Reader.* New York; London: Routledge, pp. 1–11.

Barad, K., 2003. Posthumanist performativity: Toward an understanding of how matter comes to matter. *Signs,* 28(3), 801–31.

Barad, K., 2007. *Meeting the Universe Halfway: Quantum Physics and the Entanglement of Matter and Meaning.* Durham, NC: Duke University Press.

Barad, K., 2012. Nature's queer performativity. *Women, Gender and Research,* 2012(1–2), 25–53.

Barad, K., 2014. Diffracting diffraction: Cutting together-apart. *Parallax,* 20(3), 168–87.

Barlow, J.P., 1996. *A Declaration of the Independence of Cyberspace.* [online] Available at: https://projects.eff.org/~barlow/Declaration-Final.html [Accessed 19 February 2015].

Barreto Lemos, G., Borish, D., Cole, G.D., Ramelow, S., Lapkiewicz, R. and Zeilinger, A., 2014. Quantum imaging with undetected photons. *Nature,* 512, 409–12. [online] Available at: http://www.nature.com/nature/journal/v512/ n7515/full/nature13586.html [Accessed 1 April 2015].

Baudrillard, J., 1994 [1981]. *Simulacra and Simulation.* Translated from French by S. Glaser. Ann Arbor, MI: University of Michigan Press.

Berardi Bifo, F., 2008 [2001]. *Félix Guattari: Thought, Friendship and Visionary Cartography.* Translated and edited from Italian by G. Mecchia and C.J. Stivale. New York: Palgrave Macmillan.

Berger, J., 1972. *Ways of Seeing.* London: Penguin.

Bergson, H., 2004 [1912]. *Matter and Memory.* Translated from French by N.M. Paul and W. Scott Palmer. Mineola, NY: Dover Publications.

Bey, H., 1985. *T. A. Z. The Temporary Autonomous Zone, Ontological Anarchy, Poetic Terrorism.* [online] Available at: http://hermetic.com/bey/taz_cont.html [Accessed 3 March 2015].

Bhabha, H.K., 1994. *The Location of Culture.* New York; London: Routledge.

Bhabha, H.K., 1999. Preface. Arrivals and departures. In: H. Naficy, ed. *Home, Exile, Homeland: Film, Media and the Politics of Place.* New York; London: Routledge, pp. vii–xii.

Bickmore, T.W. and Picard, R.W., 2005. Establishing and maintaining long-term human-computer relationships. *Transactions on Computer Human Interaction,* 12(2), 293–327.

Biemann, U., 1997. Outsourcing and subcontracting. Curatorial practice in post-colonial sites. In: *Kültür, ein Gender-Projekt aus IstaNbul.* Zürich: Shedhalle

Verlag. [online] Available at: http://www.geobodies.org/books-and-texts/ kultur [Accessed 19 February 2015].

Biemann, U., 2003. Geography and the politics of mobility. Introduction. In: U. Biemann, ed. *Geography and the Politics of Mobility.* Vienna: Generali Foundation. [online] Available at: http://www.geobodies.org/books-and-texts/ geography-and-the-politics-of-mobility [Accessed 19 February 2015].

Biemann, U., 2005. Remotely sensed. A topography of the global sex trade. *Feminist Review,* LXX(1), 180–93.

Biemann, U., 2007. Videographies of navigating geobodies. In: K. Marciniak, A. Imre and Á. O'Healy, eds. *Transnational Feminism in Film and Media.* New York: Palgrave Macmillan, pp. 129–45.

Biemann, U., 2008. Dispersing the viewpoint. Sahara Chronicle. In: U. Biemann and J-E. Lundström, eds. *Mission Reports: Artistic Practice in the Field: Video Works 1998–2008.* Bristol: Arnolfini Gallery Ltd, pp. 77–92.

The Black Sea Files, 2005a [video essay] Ursula Biemann. Brussels: Argos Library.

Bleecker, J. and Knowlton, J., 2006. Locative media: A brief bibliography And taxonomy of GPS-enabled locative media. *Leonardo Electronic Almanac,* 14(3). [online] Available at: http://www.leoalmanac.org/wp-content/ uploads/2012/07/Locative-Media-A-Brief-Bibliography-And-Taxonomy-Of-Gps-Enabled-Locative-Media-Vol-14-No-3-July-2006-Leonardo-Electronic-Almanac.pdf [Accessed February 23 2015].

Boccia Artieri, G., 2004. *I Media-mondo.* Rome: Meltemi.

Bohr, N., 1958. *Atom Physics and Human Knowledge.* New York: Wiley.

Bolt, B., 2004. *Art Beyond Representation: The Performative Power of the Image.* London: Tauris and Co.

Bolter, J.D. and Gromala, D., 2003. *Windows and Mirrors: Interaction Design, Digital Art and the Myth of Transparency.* Cambridge, MA: The MIT Press.

Bolter, J.D. and Grusin, R.A., 1999. *Remediation: Understanding New Media.* Cambridge, MA: The MIT Press.

Boundas, C.V., 1996. Deleuze–Bergson: An ontology of the virtual. In: P. Patton, ed. *Deleuze: A Critical Reader.* Oxford: Blackwell, pp. 80–106.

Bourdieu, P., 1997 [1972]. *Outline of a Theory of Practice.* Translated from French by R. Nice. Cambridge: Cambridge University Press.

Brah, A., 1996. *Cartographies of Diaspora: Contesting Identities.* New York: Routledge.

Braidotti, R., 1994. *Nomadic Subjects: Embodiment and Sexual Difference in Contemporary Feminist Theory.* New York: Columbia University Press.

Braidotti, R., 1996. *Madri, mostri e macchine* A.M. Crispino, ed., Roma: Manifestolibri.

Braidotti, R., 1998. *Difference, Diversity and Nomadic Subjectivity.* [online] Available at: http://www.translatum.gr/forum/index.php?topic=14317.0 [Accessed 19 February 2015].

Braidotti, R., 2002. *Metamorphoses: Towards a Materialist Theory of Becoming.* Oxford: Blackwell.

Braidotti, R., 2003. Becoming woman: Or sexual difference revisited. *Theory, Culture & Society*, 20(3), 43–64.

Braidotti, R., 2006. *Transpositions: On Nomadic Ethics*. Cambridge: Polity Press.

Braidotti, R., 2007. Feminist epistemology after postmodernism: Critiquing science, technology and globalisation. *Interdisciplinary Science Reviews*, 32(1), 65–74.

Bratton, B., 2013. *On Apps and Elementary Forms of Interfacial Life: Object, Image, superimposition*. [online] Available at: http://www.bratton.info/projects/texts/on-apps-and-elementary-forms-of-interfacial-life [Accessed 20 February 2015].

Bratton, B. and Jeremijenko, N., 2008. *Suspicious Images, Latent Interfaces*. New York: The Architectural League of New York. [online] Available at: http://www.situatedtechnologies.net/?q=node/88 [Accessed 20 February 2015].

Bridle, J., 2011. *Robot flâneur*. [website] Available at: http://robotflaneur.com [Accessed 23 February 2015].

Bridle, J., 2012–. *Dronestagram*. [website] Available at: http://dronestagram.tumblr.com [Accessed 20 February 2015].

Brown, B. and Laurier, E., 2005. En-spacing technology: Some thoughts on the geographical nature of technology. In: P. Turner and E. Davenport, eds. *Space, Spatiality and Technology*. Dordrecht: Springer, pp. 19–30.

Bruno, G., 2002. *Atlas of Emotions. Journeys in Art, Architecture and Film*. London: Verso.

Buchanan, I., 2005. Space in the age of non-place. In: I. Buchanan and G. Lambert, eds. *Deleuze and Space*. Toronto: University of Toronto Press, pp. 16–35.

Butler, J., 1990. *Gender Trouble: Feminism and the Subversion of Identity*. New York: Routledge.

Butler, J., 1997. *Excitable Speech: A Politics of the Performative*. New York: Routledge.

CAE, 2000. Recombinant theater and digital resistance. *The Drama Review*, 44, 151–66.

CAE, 2002. Introduction. Contestational biology. In: *Molecular Invasion*. New York: Autonomedia, pp. 1–14. [online] Available at: http://www.critical-art.net/books/molecular/intro.pdf [Accessed 25 February 2015].

CAE and da Costa, B,. 2001–2003. *GenTerra* [performance and website] Available at: http://www.critical-art.net/Biotech.html [Accessed 25 February 2015].

CAE, da Costa, B. and Pentecost, C., 2002–2004. *Molecular Invasion* [performance and website] Available at: http://www.critical-art.net/Biotech.html [Accessed 25 February 2015].

CAE, da Costa, B. and Shyu, S., 2003–2004. *Free Range Grains* [performance and website] Available at: http://www.critical-art.net/Biotech.html [Accessed 25 February 2015].

Caillois, R., 1984 [1935]. Mimicry and legendary psychastenia. Translated from French by J. Shepley. *October*, 31, 12–34.

Campbell, K., 2004. The promise of feminist reflexivities: Developing Donna Haraway's project for feminist science studies. *Hypatia*, 19, 162–82.

Carr, B., 2004. Universalism's irrational outburst. *Culture Machine*, 6. [online] Available at: http://www.culturemachine.net/index.php/cm/article/viewArticle/11/10 [Accessed 11 February 2015].

Carrillo Rowe, A., 2005. Be longing: Toward a feminist politics of relation. *NWSA*, 17(2), 15–46.

Casagrande, M., 2010. Urban acupuncture. [online] Available at: http://thirdgenerationcity.pbworks.com/f/urban acupuncture.pdf [Accessed 24 February 2015].

Casey, E., 1997. *The Fate of Place: A Philosophical History.* Berkeley: University of California Press.

Casey, E., 2001. Between geography and philosophy: What does it mean to be in a place world? *Annals of the Association of the American Geographers*, 91(4), 683–93.

Castells, M., 1996. *The Rise of the Network Society.* Oxford: Blackwell.

Castoriadis, C., 1994. Radical imagination and the social instituting imaginary. In: G. Robinson and J. Rundell, eds. *Rethinking Imagination: Culture and Creativity.* New York: Routledge, pp. 136–54.

Cerda Seguel, D., 2009. Land, meaning and territory: The geosemantic equation. *Inclusiva-net #2*, 10–24. [online] Available at: medialab-prado.es/mmedia/5228 [Accessed 20 February 2015].

Chicago, J. et al. 1972. *Womanhouse* [temporary multimedia installation] (Los Angeles).

Chow, R., 2006. *The Age of the World Target.* Durham, NC: Duke University Press.

Ciccoricco, D., 2004. Network vistas: Folding the cognitive map. *Image [and] Narrative*, 8. [online] Available at: http://www.imageandnarrative.be/inarchive/issue08/daveciccoricco.htm [Accessed 11 February 2015].

Clynes, M.E. and Kline, N.S., 1995 [1960]. Cyborgs and space. In: C. Habels Gray, ed. *The Cyborg Handbook.* New York: Routledge, pp. 29–33.

Code, L., 2006. *Ecological Thinking: The Politics of Epistemic Location.* Oxford: Oxford University Press.

Colebrook, C., 2005. The space of man: On the specificity of affect in Deleuze and Guattari. In: I. Buchanan and G. Lambert, eds. *Deleuze and Space.* Toronto: University of Toronto Press, pp. 189–206.

Coyne, R., 1999. *Technoromanticism: Digital Narrative, Holism, and the Romance of the Real.* Cambridge, MA: The MIT Press.

Crampton, J., 2009. Cartography: Map. 2.0. *Progress in Human Geography*, 33(1), 91–100.

Crampton, J. and Krygier, J.B., 2006. An introduction to critical cartography. *ACME Journal*, 4(1), 11–33.

Crandall, J., 2006. Precision + guided + seeing. *C-Theory.* [online] Available at: http://www.ctheory.net/articles.aspx?id=502 [Accessed 20 February 2015].

Crow, B., Longford, M., Sawchuk, K. and Zeffiro, A., 2008. Voices from beyond: Ephemeral histories, locative media and the volatile interface. In: M. Foth, ed. *Urban Informatics: The Practice and Promise of the Real-time City.* Hershey, PA: IGI Global, pp. 158–78.

da Costa, B., 2006–2008. *Pigeon Blog.* [locative media artwork] Available at: http://nideffer.net/shaniweb/pigeonblog.php [Accessed 25 February 2015].

da Costa, B., 2008. Reaching the limit: When art becomes science. In: *Tactical Biopolitics: Art, Activism and Technoscience.* Cambridge, MA: The MIT Press, pp. 365–85.

da Costa, B., 2008–2009. *Invisible Earthlings.* [workshop] Available at: http://nideffer.net/shaniweb/invisible.php [Accessed 25 February 2015].

da Costa, B., 2009. *A Memorial for the Still Living.* [multimedia installation] John Hansard Gallery, Southampton, UK.

da Costa, B., 2009–2012. *The Cost of Life.* [multimedia project] Available at: http://www.creative-capital.org/projects/view/311 [Accessed 25 February 2015].

da Costa, B., 2012. *Dying for the Other.* [video installation] Laguna Art Museum, Laguna Beach, CA, 2013.

de Certeau, M., 1984 [1980]. *The Practice of Everyday Life.* Translated from French by S. Rendall. Berkeley, CA: University of California Press.

De Lauretis, T., 1987. *Technologies of Gender: Essays on Theory, Film, and Fiction.* Bloomington, IN: Indiana University Press.

de Souza e Silva, A., 2006. From cyber to hybrid: Mobile technologies as interfaces of hybrid spaces. *Space & Culture,* 9(3), 261–78.

de Souza e Silva, A. and Soutko, D., 2011. Theorizing locative technologies through philosophies of the virtual. *Communication Theory,* 21, 23–42.

Debord, G., 1998 [1958]. Teoria della deriva. Translated from French by P. Stanziale. In: P. Stanziale, ed. *Situazionismo. Materiali per un'economia politica dell'immaginario. I testi della Internationale Situationniste.* Bolsena: Massari editore, pp. 56–63.

Del Casino, V.J., Jr., and Hanna, S.P., 2006. Beyond the 'binaries': A methodological intervention for interrogating maps as representational practices. *ACME,* IV(1), 34–56.

Deleuze, G., 1992 [1990]. Postscript on the societies of control. *October,* 59, 3–7.

Deleuze, G., 1993 [1988]. *The Fold: Leibniz and the Baroque.* Translated from French by T. Conley. Minneapolis, MN: University of Minnesota Press.

Deleuze, G., 1994 [1968]. *Difference and Repetition.* Translated from French by P. Patton. New York: Columbia University Press.

Deleuze, G. and Canguilhem, G., 2006. *Il significato della vita.* Milan: Mimesis.

Deleuze, G. and Guattari, F., 1987 [1980]. *A Thousand Plateaus: Capitalism and Schizophrenia.* Translated from French by B. Massumi. Minneapolis, MN: University of Minnesota Press.

Deutsche, R., 1995. Surprising geography. *Annals of the Association of American Geographers,* 85(1), 168–75.

Di Corinto, A. and Tozzi, T., 2006. *Hactivism. La libertà nelle maglie della rete.* Rome: Manifestolibri. [online] Available at: http://www.hackerart.org/storia/hacktivism.htm [Accessed 23 February 2015].

Dodes, R., 2013. Spike Jonze on Scarlett Johansson and 'Her'. *The Wall Street Journal.* [online] 12 December. Available at: http://www.wsj.com/news/articles/SB10001424052702304744304579250212427475406 [Accessed 14 February 2015].

Doel, A., 2010. Representation and difference. In: B. Anderson and P. Harrison, eds. *Taking Place: Non-representational Theories and Geography.* Farnham: Ashgate, pp. 117–30.

Dominguez, R., 2013. *Untitled* [video online] Available at: https://vimeo.com/24771347 [Accessed 21 March 2015].

Duncan, C., 1982. Happy mothers and other new ideas in eighteenth century art. In: N. Broude and M.D. Garrard, eds. *Feminism and Art History: Questioning the Litany.* New York: Harper & Row, pp. 201–19.

Duncan, J.S., 1993. Sites of representation. Place, time and the discourse of the other. In: J.S. Duncan and D. Ley, eds. *Place/Culture/Representation.* New York: Routledge, pp. 39–56.

EDT, 2007–. *Transborder Immigrant Tool* [locative media artwork] University of California, San Diego.

EDT, 2009. *The Transborder Immigrant Tool: Violence, Solidarity and Hope in post-NAFTA Circuits of Bodies *electr(on)/ic*.* [online] Available at: http://www.uni-siegen.de/locatingmedia/workshops/mobilehci/cardenas_the_transborder_immigrant_tool.pdf [Accessed February 2015].

Eglash, R., 2011. Multiple objectivity: An anti-relativist approach to situated knowledge. *Kybernetes,* 40(7–8), 995–1003.

Elahi, H.M., n.d. *Artist statement.* [online] Available at http://allourlives.wordpress.com/trackingtransiencenet-by-hasan-elahi [Accessed 23 February 2015].

Elahi, H.M., 2006–. *Tracking Transience* [website] Available at http://elahi.umd.edu/track [Accessed 23 February 2015].

Elovaara, P. and Mörtberg, C., 2007. Design of digital democracies: Performances of citizenship, gender and IT. *Information, Communication & Society,* 10(3), 404–23.

Escobar, A., Hess, D., Licha, I., Sibley, W., Strathern, M. and Sutz, J., 1994. Welcome to Cyberia: Notes on the anthropology of cyberculture (and comments and reply). *Current Anthropology,* 35(3), 211–31.

Europlex. 2003 [video essay] Ursula Biemann. Brussels: Argos Video Library.

Fabian, J., 1983. *Time and the Other: How Anthropology Makes its object.* New York: Columbia University Press.

Farinelli, F., 2003. *Geografia: Un'introduzione ai modelli del mondo.* Turin: Einaudi.

Farinelli, F., 2009. *La crisi della ragione cartografica.* Turin: Einaudi.

Faulkner, W., 2001. The technology question in feminism – a view from feminist technology studies. *Women's Studies International Forum,* 24(1), 79–95.

Featherstone, D., 2007. The spatial politics of the past unbound: Transnational networks and the making of political identities. *Global Networks*, 7(4), 430–52.

Featherstone, D., Phillips, R. and Waters, J., 2007. Introduction: Spatialities of transnational networks. *Global Networks*, 7(4), 383–91.

Featherstone, M., 1993. Global and local cultures. In: J. Bird, B. Curtis, T. Putnam and L. Tickner eds. *Mapping the Futures: Local Cultures, Global Change.* New York: Routledge, pp. 169–83.

FemTechNet, 2013. *Feminism, Technology and Transformation. A Live Discussion on the Life and Work of Beatriz da Costa with Donna Haraway, Catherine Lord and Alexandra Juhasz.* Laguna Art Museum, Los Angeles. [online video] Available at: https://vimeo.com/80248724 [Accessed 25 February 2015].

Fernandez, M. and Wilding, F., 2002. Situating cyberfeminisms. In: M. Fernandez, F. Wilding and M.M. Wright, eds. *Domain Errors!: Cyberfeminist Practices.* New York: Autonomedia, pp. 17–28.

Flaubert, G., 1983. *Flaubert in Egypt: A Sensibility on Tour: A narrative Drawn from Gustave Flaubert's Travel Notes & Letters.* Edited and translated from French by F. Steegmuller. London: Penguin.

Foucault, M., 1977 [1975] *Discipline and Punish: The Birth of the Prison.* Translated from French by A. Sheridan. London: Penguin.

Foucault, M., 1979 [1976]. *The History of Sexuality, vol. 1. The Will to Knowledge.* Translated from French by R. Hurley. London: Allen Lane.

Foucault, M., 1983 [1973]. *This is not a pipe.* Edited and translated from French by J. Harkness. Berkeley: University of California Press.

Foucault, M., 1986 [1984]. Of other spaces. Translated from French by J. Miskoviec. *Diacritics*, 16(1), 22–7.

Foucault, M., 1994 [1966]. *The Order of Things: An Archeology of Human Sciences.* Translated from French by A. Sheridan. New York: Vintage Books.

Foucault, M., 2003 [1963]. *The Birth of the Clinic: An Archeology of Medical Perception.* Translated from French by A. Sheridan. New York: Routledge.

Foucault, M., 2007a [1980]. Questions on geography. Translated from French by C. Gordon. In: J.W. Crampton and S. Elden, eds. *Space, Knowledge and Power: Foucault and Geography.* Farnham: Ashgate, pp. 172–82.

Foucault, M., 2007b [1978]. The incorporation of the hospital into modern technology. Translated from French by E. Knowlton, Jr., W.J. King and S. Elden. In: J.W. Crampton and S. Elden, eds. *Space, Knowledge and Power: Foucault and Geography.* Farnham: Ashgate, pp. 141–52.

Franklin, S., Lury, C. and Stacey, J., 2000. *Global Nature, Global Culture.* London: Sage.

Friedberg, A., 1993. *Window Shopping: Cinema and the Postmodern.* Berkeley, CA: University of California Press.

Friedman, S., 1998. *Mappings: Feminism and the Cultural Geographies of Encounter.* Princeton, NJ: Princeton University Press.

Fuller, M., 2005. *Media Ecologies: Materialist Energies in Art and Technoculture.* Cambridge, MA: The MIT Press.

Fusco, C., 2004. Questioning the frame. Thoughts about maps and spatial logic in the global present. *Inthesetimes.com*. [online] 16 December. Available at: http://www.inthesetimes.com/site/main/article/1750 [Accessed 23 February 2015].

Fusco, C. and Wallis, B., 2003. *Only Skin Deep: Changing Visions of the American Self*. New York: International Center of Photography, Harry N. Abrams.

Gadamer, H-G., 1975 [1960]. *Truth and Method*. Translated from German by G. Barden and J. Cumming. London: Sheed and Ward.

Gajjala, R. and Mamidipudi, A., 2002. Gendering processes within technological environments: A cyberfeminist issue. *Rhizomes*, 4(Spring). [online] Available at: http://www.rhizomes.net/issue4/gajjala.html [Accessed 19 February 2015].

Galloway, A.R., 2012. *The Interface Effect*. Cambridge UK: Polity Press.

Gane, N., 2006. When we have never been human, what is to be done? Interview with Donna Haraway. *Theory Culture & Society*, 23(135), 135–58.

García Selgas, F.J., 2004. Feminist epistemologies for critical social theory: From standpoint theory to situated knowledge. In: S. Harding, ed. *The Feminist Standpoint Theory Reader: Intellectual & Political Controversies*. New York: Routledge, pp. 293–308.

Gemini, L., 2008. *In viaggio. Immaginario, comunicazione e pratiche del turismo contemporaneo*. Milan: Franco Angeli.

Ghani, A., 1993. Space as an arena of represented practices. In: J. Bird, B. Curtis, T. Putnam and L. Tickner eds. *Mapping the Futures: Local Cultures, Global Change*. New York: Routledge, pp. 47–58.

Gordon, E., 2007. Mapping digital networks from cyberspace to Google. *Information, Communication & Society*, 10(6), 885–901.

Gordon, E., 2010. *The Urban Spectator: American Concept-cities from Kodak to Google*. Lebanon, NH: Dartmouth College Press.

Gordon, E. and de Souza e Silva, A., 2011. *Net Locality: Why Location Matters in a Networked World*. Malden, MA: Wiley-Blackwell.

Gough, N., 1994. Narration, reflection, diffraction: aspects of fiction in educational inquiry. *The Australian Educational Researcher*, 21(3), 47–76.

Graham, S., 1998. The end of geography or the explosion of place? Conceptualizing space, place and information technology. *Progress in Human Geography*, 22(2), 165–85.

Graham, S., 2004a. Introduction: From dreams of transcendence to the remediation of urban life. In: S. Graham, ed. *The Cybercities Reader*. New York: Routledge, pp. 1–30.

Graham, S., 2004b. Beyond the 'dazzling light': From dreams of transcendence to the 'remediation' of urban life: A research manifesto. *New Media & Society*, 6(1), 16–25.

Gregory, D., 1994. *Geographical Imaginations*. Oxford: Blackwell.

Grewal, I. and Kaplan, C., 1994. Introduction. Transnational feminist practices and questions of postmodernity. In: I. Grewal and C. Kaplan, eds. *Scattered Hegemonies: Postmodernity and Transnational Feminist Practices*. Minneapolis, MN: University of Minnesota Press, pp. 1–33.

Grossberg, L., 1996. The space of culture, the power of space. In: I. Chambers and L. Curti, eds. *The Post-colonial Question*. New York: Routledge, pp. 169–88.

Grossman, R., 1980. Women's place in the integrated circuit. *Radical America*. 14(1), 29–50.

Grosz, E., 2008. *Chaos, Art, Territory: Deleuze and the Framing of Earth*. New York: Columbia University Press.

Grusin, R., 2000. Location, location, location. Desktop real estate and the cultural economy of the World Wide Web. *Convergence*, 6(1), 48–61.

Grusin, R., 2010. *Premediation: Affect and Mediality after 9/11*. New York: Palgrave Macmillan.

Guattari, F., 1984 [1977]. *Molecular Revolution: Psychiatry and Politics*. Translated from French by R. Sheed. London: Penguin.

Guattari, F., 1995 [1992]. *Chaosmosis: An Ethico-aesthetic Paradigm*. Translated from French by P. Bains and J. Pefanis. Sydney: Power Publications.

Guattari, F., 2000 [1989]. *The Three Ecologies*. Translated from French by I. Pindar and P. Sutton. London: The Athlone Press.

Guattari, F., 2013 [1989]. *Schizoanalytic Cartographies*. Translated from French by A. Goffey. London: Bloomsbury Publishing.

Haden, D., 2008. *A Short Enquiry into the Origins and Use of the Term Neogeography*. [online] Available at: http://www.d-log.info/on-neogeography. pdf [Accessed 20 February 2015].

Hansen, M.B.N., 2006. *Bodies in Code: Interfaces with Digital Media*. New York: Routledge.

Haraway, D., 1991. *Simians, Cyborgs and Women: The Reinvention of nature*. New York: Routledge.

Haraway, D., 1992. The promises of monsters: A regenerative politics for inappropriate/d others. In: L. Grossberg, C. Nelson and P.A. Treichler, eds. *Cultural studies*. New York: Routledge, pp. 295–337.

Haraway, D., 1997. *Modest_Witness@Second_Millennium. FemaleMan©_ Meets_OncoMouse™: Feminism and Technoscience*. New York: Routledge.

Haraway, D., 2000. *How Like a Leaf: An Interview with Thyrza Nichols Goodeve*. New York: Routledge.

Haraway, D., 2003. *The Companion Species Manifesto: Dogs, People and Significant Otherness*. Chicago, IL: Prickly Paradigm Press.

Haraway, D., 2008a. Otherwordly conversations, terrain topics, local terms. In: S. Alamo and S. Hekman, eds. *Material Feminisms*. Bloomington, IN: Indiana University Press, pp. 157–87.

Haraway, D., 2008b. *When Species Meet*. Minneapolis, MN: University of Minnesota Press.

Harding, S., 1986. *The Science Question in Feminism*. Ithaca, NY: Cornell University Press.

Harding, S., 1991. Strong objectivity and socially situated knowledge. In: *Whose Science? Whose Knowledge?* Ithaca, NY: Cornell University Press, pp. 138–63.

Harding, S., 2003. Representing reality. The critical realism project. *Feminist Economics*, 9(1), 151–59.

Harding, S., 2004a. Introduction: Standpoint theory as a site of political, philosophic, and scientific debate. In: S. Harding, ed. *The Feminist Standpoint Theory Reader: Intellectual & Political Controversies*. New York: Routledge, pp. 1–15.

Harding, S., 2004b [1997]. Comment on Hekman's 'Truth and method: Feminist standpoint theory revisited': Whose standpoint needs the regimes of truth and reality? In: S. Harding, ed. *The Feminist Standpoint Theory Reader: Intellectual & Political Controversies*. New York: Routledge, pp. 255–62.

Harding, S., 2004c [1993]. Rethinking standpoint epistemology: What is 'strong objectivity'? In: S. Harding, ed. *The Feminist Standpoint Theory Reader: Intellectual & Political Controversies*. New York: Routledge, pp. 127–42.

Harding, S., 2008. *Sciences from Below: Feminisms, Postcolonialities and Modernities*. Durham, NC: Duke University Press.

Hardt, M. and Negri, A., 2000. *Empire*. Cambridge, MA: Harvard University Press.

Hartsock, N., 2004 [1983]. The feminist standpoint: Developing the ground for a specifically feminist historical materialism. In: S. Harding, ed. *The Feminist Standpoint Theory Reader: Intellectual & Political Controversies*. New York: Routledge, pp. 35–53.

Harvey, D., 1989. *The Condition of Postmodernity: An Enquiry into the Origins of Cultural Change*, Oxford: Blackwell.

Harvey, D., 1993. From space to place and back again: Reflections on the condition of postmodernity. In: J. Bird, B. Curtis, T. Putnam and L. Tickner eds. *Mapping the Futures: Local Cultures, Global Change*. New York: Routledge, pp. 2–29.

Harvey, D., 2007. The Kantian roots of Foucault's dilemmas. In: J. Crampton and S. Elden, eds. *Space, Knowledge and Power: Foucault and Geography*. Farnham: Ashgate, pp. 41–8.

Hauser, J., 2008. Observations on an art of growing interest: toward a phenomenological approach to art involving biotechnology. In: B. da Costa and K. Philip, eds. *Tactical Biopolitics: Art, Activism and Technoscience*. Cambridge, MA: The MIT Press, pp. 83–104.

Hayles, K., 1995. Making the cut: The interplay of narrative and system, or what systems theory can't see. *Cultural Critique*, 30(Spring), 71–100.

Hayles, K., 1997 [1993]. *Constrained Constructivism: Locating Scientific Inquiry in the Theater of Representation*. [online] Part 1 available at: http://www. nettime.org/Lists-Archives/nettime-l-9704/msg00037.html; part 2 available at: http://www.nettime.org/Lists-Archives/nettime-l-9704/msg00038.html [Accessed 14 February 2015].

Hayles, K., 1999. *How We Became Posthuman: Virtual Bodies in Cybernetics, Literature, and Informatics*. Chicago, IL: University of Chicago Press.

Hayles, K., 2005. *My Mother was a Computer: Digital Subjects and Literary Texts*. Chicago, IL: The University of Chicago Press.

Hayles, K., Luhmann, N., Rasch, W., Knodt, E. and Wolfe, C., 1995. Theory of a different order: A conversation with Katherine Hayles and Niklas Luhmann. *Cultural Critique*, 31(Fall), 7–36.

Heidegger, M., 1962 [1927]. *Being and Time*. Translated from German by J. Macquarrie and E. Robinson. New York: Harper & Row.

Hemment, D., 2004. The locative dystopia. *nettime.org*. [online] Available at: http://eprints.lancs.ac.uk/30831 [Accessed 23 February 2015].

Hemment, D., 2006. Locative arts. *Leonardo*, 39(4), 348–55.

Her, 2013.[Film] Directed by Spike Jonze. USA: Annapurna Pictures.

Hill Collins, P., 2004 [1986]. Learning from the outsider within: The sociological significance of black feminist thought. In: S. Harding, ed. *The Feminist Standpoint Theory Reader: Intellectual & Political Controversies*. New York: Routledge, pp. 103–26.

Hirschmann, N.J., 2004 [1997]. Feminist standpoint as postmodern strategy. In: S. Harding, ed. *The Feminist Standpoint Theory Reader: Intellectual & Political Controversies*. New York: Routledge, pp. 317–32.

Holmes, B., 2000. *Imaginary Maps, Global Solidarities*. [online] Available at: https://brianholmes.wordpress.com/2000/10/20/imaginary-maps-global-solidarities [Accessed 11 February 2015].

Holmes, B., 2009. Guattari's schizoanalytic cartographies or, the pathic core at the heart of cybernetics. *Deriva Continental* [blog] 27 February. Available at: https://brianholmes.wordpress.com/2009/02/27/guattaris-schizoanalytic-cartographies [Accessed 25 February 2015].

hooks, b., 1990. *Yearning: Race, Gender, and Cultural Politics*. Boston, MA: South End Press.

Huber, J., 2003. Video-essayism. On the theory-practice of the transitional. In: U. Biemann, ed. *Stuff it: The Video-essay in the Digital Age*. Vienna: Springer, pp. 92–7.

Huber, J., 2008. Getting to the bottom of vision: Theory of images, images of theory. In: U. Biemann and J-E. Lundström, eds. *Mission Reports: Artistic Practice in the Field: Video Works 1998–2008*. Bristol: Arnolfini Gallery, pp. 170–77.

Hughes, C. and Lury, C., 2013. Re-turning feminist methodologies: From a social to an ecological epistemology. *Gender and Education*, 25(6), 786–99.

Iaconesi, S. and Persico, O., 2014. Urban acupuncture in the era on ubiquitous media. *The Journal of Community Informatics*, 10(3). [online] Available at: http://ci-journal.net/index.php/ciej/article/view/1073 [Accessed 23 February 2015].

Iaconesi, S. and Persico, O., 2015. *La cura. Working Book*. [unpublished manuscript].

Irigaray, L., 1991. *The Irigaray Reader*. M. Whitford, ed., Oxford: Blackwell.

Ito, M., 1999. Network localities: Identity, place, and digital media. *Meetings of the Society for the Social Studies of Science*, San Diego. [online] Available at: www.itofisher.com/PEOPLE/mito/locality.pdf [Accessed 18 February 2015].

Jacobs, J. and Nash, C., 2003. Too little, too much: Cultural feminist geographies. *Gender, Place and Culture – A Journal of Feminist Geography*, 10(3), 265–79.

Jameson, F., 1988. Cognitive mapping. In: C. Nelson and L. Grossberg, eds. *Marxism and the Interpretation of Culture*. Urbana, IL: University of Illinois Press, pp. 347–60.

Jameson, F., 1991. *Postmodernism, or, the Cultural Logic of Late Capitalism*. Durham, NC: Duke University Press.

Jeremijenko, N., n.d. *The Environmental Health Clinic* [website] Available at: http://www.environmentalhealthclinic.net [Accessed 23 February 2015].

Jones, A. ed., 1996. *Judy Chicago's Dinner Party in Feminist Art History*. Berkeley: University of California Press.

Kaplan, C., 1994. The politics of location as transnational feminist critical practice. In: I. Grewal and C. Kaplan, eds. *Scattered Hegemonies: Postmodernity and Transnational Feminist Practices*. Minneapolis, MN: University of Minnesota Press, pp. 137–52.

Kaplan, C., 2002. Transporting the subject: Technologies of mobility and location in an era of globalization. *PMLA*, 117(1), 32–42.

Karentzos, A., 2013. Lisl Ponger's passages – In-between tourism and migration. In: S. Witzgall, G. Vogl and S. Kesserling, eds. *New Mobilities Regimes in Art and Social Sciences*. Farnham: Ashgate, pp. 135–47.

Kember, S., 2003. *Cyberfeminism and Artificial Life*. New York: Routledge.

Kember, S. and Zylinska, J., 2012. *Life after New Media: Mediation as a Vital Process*. Cambridge, MA: The MIT Press.

Kesserling, S. and Vogl, G., 2013. The new mobilities regimes. In: S. Witzgall, G. Vogl and S. Kesserling, eds. *New Mobilities Regimes in Art and Social Sciences*. Farnham: Ashgate, pp. 17–36.

Kingsbury, P. and Jones, J.P., 2009. Walter Benjamin's Dionysian adventures on Google Earth. *Geoforum*, 40(4), 502–13.

Kirby, K., 1996. Re: Mapping subjectivity. Cartographic vision and the limits of politics. In: N. Duncan, ed. *BodySpace: Destabilising Geographies of Gender and Sexuality*. New York: Routledge, pp. 45–55.

Kitchin, R. and Dodge, M., 2011. *Code/Space: Software and Everyday Life*. Cambridge, MA: The MIT Press.

Kitchin, R., Perkins, C.R. and Dodge, M., 2009. Thinking about maps. In: R. Kitchin, C.R. Perkins and M. Dodge, eds. *Rethinking Maps*. New York: Routledge, pp. 1–25.

Knorr Cetina, K., 2001. Postsocial relations: Theorizing sociality in a postsocial environment. In: G. Ritzer and B. Smart, eds. *Handbook of Social Theory*. London: Sage, pp. 520–37.

Kotànyi, A. and Vaneigem, R., 1998 [1961]. Programma elementare dell'ufficio di urbanismo unitario. Translated from French by P. Stanziale. In: P. Stanziale, ed. *Situazionismo. Materiali per un'economia Politica dell'immaginario. I testi della Internationale Situationniste*. Bolsena: Massari editore, pp. 73–7.

Krauss, R., 1985 [1977]. Grids. In: *The Originality of the Avant-garde and Other Modernist Myths*. Cambridge, MA: The MIT Press, pp. 9–22.

Kroker, A., 2010. Minor simulations, major disturbances. *Empyre Mailing List*. [online] 13 April. Available at: http://lists.cofa.unsw.edu.au/pipermail/empyre/2010-April/002825.html [Accessed 23 February 2015].

Kwan, M-P., 2002. Feminist visualization: Re-envisioning GIS as a method in feminist geographic research. *Annals of the Association of American Geographers*, 92(4), 645–61.

Kwan, M-P., 2004. Beyond difference: From canonical geography to hybrid geographies. *Annals of the Association of American Geographers*, 94(4), 756–63.

Kwan, M-P., 2007. Affecting geospatial technologies: Toward a feminist politics of emotion. *The Professional Geographer*, 59(1), 22–34.

Lacan, J., 2007. *Ecrits: The First Complete Edition in English*. Translated from French by B. Fink. New York: Norton and Company.

Laclau, E., 1990. *New Reflections on the Revolution of our time*. London: Verso.

Larson, J., 2013. Her and the complex legacy of the female robot. *The Atlantic*. [online] Available at: http://m.theatlantic.com/entertainment/archive/2013/12/-em-her-em-and-the-complex-legacy-of-the-female-robot/282581 [Accessed 15 February 2015].

Latour, B., 1987. *Science in Action*. Cambridge, MA: Harvard University Press.

Latour, B., 1993 [1991]. *We have Never Been Modern*. Translated from French by C. Porter. Cambridge, MA: Harvard University Press.

Latour, B., 1994. On technical mediation – Philosophy, sociology, genealogy. *Common Knowledge*, 3(2), 29–64.

Latour, B., 2004. Why has critique run out of steam? From matters of fact to matters of concern. *Critical Inquiry*, 30(Winter), 225–48.

Latour, B., 2005. *Reassembling the Social: An Introduction to Actor-Network-Theory*. New York: Oxford University Press.

Latour, B., 2010 [1996]. *On the Modern Cult of the Factish Gods*. Translated from French by H. MacLean and C. Porter. Durham, NC: Duke University Press.

Lazzarato, M., 2006. The concepts of life and the living in the societies of control. In: M. Fugsland and B.M. Sorensen, eds. *Deleuze and the Social*. Edinburgh: Edinburgh University Press, pp. 171–90.

Lefebvre, H., 1991 [1974]. *The Production of Space*. Translated from French by D. Nicholson-Smith. Oxford: Blackwell.

Leibniz, G., 1956. Metaphysical foundations of mathematics. L. Loemker, ed. *Philosophical Papers and Letters*. Chicago, IL: University of Chicago Press.

Leibniz, G., 1973. *Philosophical Writings*. G.H.R. Parkinson, ed. London: Dent.

Lemos, A., 2008. Mobile communication and new sense of places: A critique of spatialization in cyberculture. *Revista Galáxia*, 16, 91–108. [online] Available at: http://revistas.pucsp.br/index.php/galaxia/article/viewPDFInterstitial/1914/1177 [Accessed 19 February 2015].

Lemos, A., 2009. Locative media manifesto. *404NotFound*, 1(71). [online] Available at: http://www.andrelemos.info/2009/05/locative-media-manifesto. html [Accessed 5 April 2015].

Le-Phat Ho, S., 2008. Locative media as war. *The Thing*. [online] 9 June. Available at: http://post.thing.net/node/2201 [Accessed 19 February 2015].

Lettvin, J., Maturana, H.R., McCulloch, W.S. and Pitts W.H., 1959. What the frog's eye tells the frog's brain. *Proceedings of the Institute of Radio Engineers*, 47, 1940–59.

Lewis, T., 2015. Rise of the fembots: Why artificial intelligence is often female. *Livescience*. [online] Available at: http://www.livescience.com/49882-why-robots-female.html [Accessed 18 March 2015].

Lohan, M., 2000. Constructive tensions in feminist technology studies. *Social Studies of Science*, 30(6), 895–916.

The Look of Silence, 2014 [Film] Directed by Joshua Oppenheimer. USA: Final Cut for Real.

Lorimer, H., 2005. Cultural geography: The busyness of being 'more-than-representational'. *Progress in Human Geography*, 29(1), 83–94.

Luhmann, N., 1990. The cognitive program of constructivism and a reality that remains unknown. In: W. Krohn, G. Kiippers and H. Nowotny, eds. *Selforganization: Portrait of a Scientific Revolution*. Dordrecht: Kluwer, pp. 64–85.

Luhmann, N., 2000 [1995]. *Art as a Social System*. Translated from German by E.M. Knodt. Stanford: Stanford University Press.

Lury, C., Parisi, L. and Terranova, T., 2012. The becoming topological of culture. *Theory Culture and Society*, 29(4/5), 3–35.

Lussault, M. and Stock, M., 2009. 'Doing with space': Towards a pragmatics of space. *Social Geography Discussions*, 5, 1–23.

Lynch, K., 1960. *The Image of the City*. Cambridge, MA: The MIT Press.

Lynes, K.G., 2011. On the ground. Media in conflict zones. In: P. Mörtenblöck and H. Moshammer, eds. *Space (Re) solutions: Interventions and Research in Visual Culture*. Bielefeld: Transcript, pp. 17–28.

Lynes, K.G. 2013. *Prismatic Media*. New York: Palgrave Macmillan.

Magritte, R., 1936. *Les liaisons dangereuses*.[oil on canvas] (New York, *Museum of Modern Art*).

Magritte, R., 1937. *La reproduction interdite*. [oil on canvas] (Rotterdam, Boijmans Van Beuningen Museum).

Mani, L. and Frankenberg, R., 1993. Crosscurrents, crosstalk: Race, 'post-coloniality', and the politics of location. *Cultural Studies*, 7, 292–310.

MapFanIncrement P Corp. 2012. *MapFan Eye*. [online video] Available at: http://www.youtube.com/watch?v=8X4IXrRF_hU [Accessed 20 February 2015].

Marston, S.A., Jones, J.P., III and Woodward, K., 2005. Human geography without scale. *Transactions of the Institute of British Geographers*, 30, 416–32.

Massey, D., 1992. Politics and space/time. *New Left Review*, 196, 65–84.

Massey, D., 1994. A global sense of place. In: *Space, Place and Gender*. Cambridge: Polity Press, pp. 146–56.

Massey, D., 2005. *For Space*. London: Sage.

Massumi, B., 2002. *Parables for the Virtual: Movement, Affect, Sensation*. Durham, NC: Duke University Press.

Massumi, B., 2014. *What Animals Teach us about Politics*. Durham, NC: Duke University Press.

McDowell, L., 1996. Spatializing feminism. Geographic perspectives. In: N. Duncan, ed. *BodySpace: Destabilising Geographies of Gender and Sexuality*. New York: Routledge, pp. 28–44.

McGarrigle, C., 2009. The construction of locative situations: Locative media and the Situationist International, recuperation or redux? *Digital Creativity*, 21, 55–62.

McLuhan, M., 1962. *The Gutenberg Galaxy*. Toronto: University of Toronto Press.

McLuhan, M., 1964. *Understanding Media: The Extensions of Man*. New York: McGraw-Hill.

Meek, A., 2000. Exile and the electronic frontier. Critical intellectuals and cyberspace. *New Media & Society*, 2(1), 85–104.

Meyrowitz, J., 1985. *No Sense of Place: The Impact of Electronic Media on Social Behavior*. New York: Oxford University Press.

Meyrowitz, J., 2005. The Rise of glocality. New senses of place and identity in the global village. In: K. Nyiri, ed. *A Sense of Place: The Global and the Local in Mobile Communication*. Vienna: Passagen, pp. 21–30.

Mezzadra, S. and Neilson, B., 2013. *Border as Method or, the Multiplication of Labor*. Durham, NC: Duke University Press.

Minh-Ha, T.T. and Gržinić, M., 1998. *Inappropriate/d Artificiality*. [online] Available at: http://trinhminh-ha.squarespace.com/inappropriated-articificiality [Accessed 19 February 2015].

Mirrlees, T., 2005. Cognitive mapping or, the resistant element in the work of Fredric Jameson: A response to Jason Berger. *Cultural Logic: An Electronic Journal of Marxist Theory and Practice*, 8. [online] Available at: http://clogic.eserver.org/2005/mirrlees.html [Accessed 11 February 2015].

Mirror Image, 2013 [Film] Directed by Danielle Schwartz. Israel: Zochrot.

Mitchell, T., 1989. The world as exhibition. *Comparative Studies in Society and History*, 31(2), 217–36.

Mitchell, W.J.T., 2008. Addressing Media. *Mediatropes*, 1, 1–18. [online] Available at: http://www.mediatropes.com/index.php/Mediatropes/article/viewArticle/1771 [Accessed 11 February 2015].

Mitchell, W., 2000. *e-topia*, Cambridge, MA: The MIT Press.

Mohanty, C.T., 1995. Feminist encounters: Locating the politics of experience. In: L. Nicholson and S. Seidman, eds. *Social Postmodernism: Beyond Identity Politics*. Cambridge: Cambridge University Press, pp. 68–86.

Mohanty, C.T., 2003. *Feminism without Borders: Decolonizing Theory, Practicing Solidarity*. Durham, NC: Duke University Press.

Mol, A., 2002. *The Body Multiple: Ontology in Medical Practice.* Durham, NC: Duke University Press.

Moore, C., 2013. Art in the after. *Art21 Magazine.* [online] Available at: http:// blog.art21.org/2013/06/12/art-in-the-after [Accessed 25 February 2015].

Munster, A., 2006. *Materializing New Media: Embodiment in Information Aesthetics.* Lebanon, NH: Dartmouth College Press, University Press of New England.

Munster, A., 2013. *An Aesthesia of Networks: Conjunctive Experience in Art and Technology.* Cambridge, MA: The MIT Press.

Murray, T., 2008. *Digital Baroque: New Media Art and Cinematic Folds.* Minneapolis, MN: University of Minnesota Press.

Nast, H.J., 1994. Women in the field: Critical feminist methodologies and theoretical perspectives. *Professional Geographer*, 46(1), 54–66.

Nast, H.J., 1998. Unsexy geographies. *Gender, Place and Culture – A Journal of Feminist Geography*, 5(2), 191–206.

Negroponte, N., 1995. *Being Digital.* New York: Alfred A. Knopf.

Nettleton, S., 2004. The emergence of e-scaped medicine. *Sociology*, 38(4), 661–79.

Nochlin, L., 1992. Morisot's wet nurse. The construction of work and leisure in Impressionist painting. In: N. Broude and M.D. Garrard, eds. *The Expanding Discourse: Feminism and Art History.* New York: HarperCollins, pp. 231–44.

Nold, C., 2004–. *Bio Mapping.* [locative media project] Available at http:// biomapping.net [Accessed 15 April 2015].

Nold, C., 2009. *Emotional Cartography: Technologies of the Self.* [online] Available at: http://emotionalcartography.net/EmotionalCartography.pdf [Accessed 23 February 2015].

November, V., Camacho-Hübner, E. and Latour, B., 2010. Entering a risky territory: Space in the age of digital navigation. *Environment and Planning D: Society and Space*, 28, 581–99.

Oliver, J., 2012. *Border Bumping.* [locative media artwork] Available at: http:// borderbumping.net [Accessed 23 February 2015].

Oppenheimer, J., 2014. *Joshua Oppenheimer on The Look of Silence* (2014 Venice Film Festival). [online video] Available at: https://www.youtube.com/ watch?v=qUb2FBPseGk [Accessed 15 February 2015].

Pacilio, L., 2014. Lei. *Gli spietati.* [online] Available at: http://www.spietati.it/z_ scheda_dett_film.asp?idFilm=5239 [Accessed 14 February 2015].

Pacteau, F., 1994. *The Symptom of Beauty.* Cambridge, MA: Harvard University Press.

Parallax, 2014. Special issue: *Diffracted Worlds – Diffractive Readings: Onto-Epistemologies and the Critical Humanities.* B.G. Kaiser and K. Thiele eds. 20(3), 165–287.

Parikka, J., 2010. *Insect Media: An Archeology of Animals and Technology.* Minneapolis, MN: University of Minnesota Press.

Parisi, L. and Terranova, T., 2000. Heat-death. Emergence and control in genetic engineering and artificial life. *C-Theory*. [online] Available at: http://www.ctheory.net/articles.aspx?id=127 [Accessed 14 February 2015].

Parker, C., 2012–2013. *Unsettled.* [multimedia installation] (London, Frith Street Gallery).

Parks, L., 2005. *Cultures in Orbit: Satellites and the Televisual*. Durham, NC: Duke University Press.

Patterson, J., 2013. Spike Jonze on Jackass, Scarlett Johansson's erotic voice and techno love. *The Guardian*. [online] Available at: http://www.theguardian.com/film/2013/nov/28/spike-jonze-her-interview-scarlett-johansson-joaquin-phoenix-jackass [Accessed 14 February 2015].

Pavlovskaya, M. and St. Martin, K., 2007. Feminism and geographic information systems: From a missing object to a mapping subject. *Geography Compass*, 1(3), 583–606.

Performing the Border. 1999 [video essay] Ursula Biemann. Brussels: Argos Video Library.

Peschl, M.F. and Riegler, A., 1999. Does representation need reality? Rethinking epistemological issues in the light of recent developments and concepts in cognitive science. In: A. Riegler, M. Peschl and A. von Stein, eds. *Understanding Representation in the Cognitive Sciences*. New York: Kluwer Academic/Plenum Publishers, pp. 9–17.

Piazza, L. Lummen, T.T., Quinoñez, E., Murooka, Y., Reed, B.W., Barwick B. and Carbone, F., 2015. Simultaneous observation of the quantization and the interference pattern of a plasmonic near-field. *Nature Communications*, 6(6207). [online] Available at: http://www.nature.com/ncomms/2015/150302/ncomms7407/full/ncomms7407.html [Accessed 1 April 2015].

Pickering, A., 1994. After representation: Science studies in the performative idiom. *PSA: Proceedings of the Biennial Meeting of the Philosophy of Science Association. Symposia and Invited Papers*, 1994(2), 413–19.

Plant, S., 1995. The future looms: Weaving women and cybernetics. *Body and Society*, 1(3–4), 45–64.

Pollock, G., 1988. Femininist interventions in the history of art. In: *Vision and Difference: Femininity, Feminism, and Histories of Art: An Introduction*. New York: Routledge, pp. 1–24.

Pope, S., 2005. The Shape of locative media. *Mute Magazine*, 1(29). [online] 9 February. Available at: http://www.metamute.org/editorial/articles/shape-locative-media [Accessed 23 February 2015].

Poster, M., 2004. Digitally local communications: Technologies and space. *The Global and the Local in Mobile Communication: Places, Images, People, Connections*. Budapest, 10–12 June. [online] Available at: http://www.phil-inst.hu/mobil/2004/poster_webversion.doc [Accessed 23 February 2015].

Proboscis, 2002–2004. *Urban Tapestries.* [locative media project and software platform] Available at http://urbantapestries.net [Accessed 15 April 2015].

Probyn, E., 1990. Travels in the postmodern: Making sense of the local. In: L.J. Nicholson, ed. *Feminism/Postmodernism*. New York: Routledge, pp. 176–89.

Prometheus, 2012 [Film] Directed by Ridley Scott. USA: Twentieth Century Fox.

Propen, A.D., 2006. Critical GPS: Toward a new politics of location. *ACME: An International E-Journal for Critical Geographies*, 4(1), 131–44.

Puig de la Bellacasa, M., 2012. 'Nothing comes without its world': Thinking with care. *The Sociological Review*, 60(2), 197–216.

Rabinow, P., 1996. *Essays on the Anthropology of Reason*. Princeton, NJ: Princeton University Press.

Rees, R., 1980. Historical links between cartography and art. *Geographical Review*, 70(1), 60–78.

Reich, L., 2014. Him: Love in the time of operating systems. *The Atlantic*. [online] Available at: http://m.theatlantic.com/technology/archive/2014/01/him-love-in-the-time-of-operating-systems/283062 [Accessed 15 February 2015].

Relph, E., 1992. Place, postmodern landscape and heterotopia. *Environmental and Architectural Phenomenology Newsletter*, 3(1), p. 14.

Remote Sensing, 2001 [video essay] Ursula Biemann. Brussels: Argos Video Library.

Rich, A., 1986. *Blood, Bread, and Poetry: Selected Prose, 1979-1985*. New York: Norton.

Robinson, J., 2000. Feminism and the spaces of transformation. *Transactions of the Institute of British Geographers*, 25(3), 285–301.

Rogers, R., 2012. Mapping and the politics of web space. *Theory, Culture & Society*, 29 (4/5), 193–219.

Rogoff, I., 2000. *Terra Infirma: Geography's Visual Culture*. New York: Routledge.

Rose, G., 1993. *Feminism and Geography: The Limits of Geographical Knowledge*. Cambridge: Polity Press.

Rose, G., 1995a. Place and identity: a sense of place. In: D. Massey and P. Jess, eds. *A Place in the World? Places, Cultures and Globalization*. Oxford: Oxford University Press.

Rose, G., 1995b. Making space for the female subject of feminism. In: S. Pile and N.J. Thrift, eds. *Mapping the Subject: Geographies of Cultural Transformation*. New York: Routledge, pp. 332–54.

Rose, G., 1996. As if the mirrors had bled. Masculinist dwelling, Masculinist theory and feminist masquerade. In: N. Duncan, ed. *BodySpace: Destabilising Geographies of Gender and Sexuality*. New York: Routledge, pp. 56–74.

Rouse, J., 2004 [1996]. Feminism and the social construction of scientific knowledge. In: S. Harding, ed. *The Feminist Standpoint Theory Reader: Intellectual & Political Controversies*. New York: Routledge, pp. 353–74.

Russell, B., 1999. Headmap manifesto. Know your place (location-aware devices). [online] Available at: http://digital.typepad.com/headmapmanifesto.pdf [Accessed 23 February 2015].

Russell, B., 2006. *Transcultural Mapping Reader (TCM) online – Introduction*, Riga: RIXC Centre for New Media Culture. [online] Available at: http://web.archive.org/web/20060613044920/locative.net/tcmreader/index.php?intro;russell [Accessed 23 February 2015].

Said, E.W., 1978. *Orientalism*. New York: Pantheon Books.

Salmon, N. and Bassett, R., 2009. Harried by Harding and Haraway: Student–mentor collaboration in disability studies. *Disability & Society*, 24(7), 911–24.

Sandoval, C., 2000. *Methodology of the Oppressed*. Minneapolis, MN: University of Minnesota Press.

Sassen, S., 1998. Towards a feminist analytics of the global economy. In: *Globalization and Its Discontents*. New York: The New York Press, pp. 81–110.

Sassen, S., 2002. Towards a sociology of information technology. *Current Sociology*, 50(3), 365–88.

Schaeffer-Grabiel, F., 2004. Cyberbrides and global imaginaries: Mexican women's turn from the national to the foreign. *Space and Culture*, 7(1), 33–48.

Schneider, J., 2005. *Donna Haraway: Live Theory*. New York: Continuum.

Schneider, R., 1997. *The Explicit Body in Performance*. New York: Routledge.

Schuurman, N. and Pratt, G., 2002. Care of the subject: Feminism and critiques of GIS. *Gender, Place and Culture – A Journal of Feminist Geography*, 9(3), 291–9.

Schwartz, D., 2014. *Password Request* [email] (Personal communication, 19 September 2014).

Serfaty, V., 2005. Cartographie d'Internet: du virtuel à la re-territorialisation. , 13, 83–96.

Shannon, E.C. and Weaver, W. 1963. *The Mathematical Theory of Communication*. Urbana, IL: University of Illinois Press.

Shepard, M. 2010–2011. *The Serendipitor.* [website] Available at: http://serendipitor.net/site [Accessed 23 February 2015].

Shields, R., 2006. Flânerie for cyborgs. *Theory, Culture & Society*, 23(7–8), 209–20.

Smith, L.C., 2012. Decolonizing hybridity: Indigenous video, knowledge, and diffraction. *Cultural Geographies*, 19, 329–48.

Smith, N. and Katz, C., 1993. Grounding metaphor: Towards a spatialized politics. In: M. Keith and S. Pile, eds. *Place and the Politics of Identity*. New York: Routledge, pp. 67–83.

Soja, E., 1989. *Postmodern Geographies: The Reassertion of Space in Critical Social Theory*. London: Verso.

Soja, E. and Hopper, B., 1993. The spaces that difference makes: Some notes on the gepgraphical margins of the new cultural politics. In: M. Keith and S. Pile, eds. *Place and the Politics of Identity*. New York: Routledge, pp. 183–205.

Solomon-Godeau, A., 1989. Going native. *Art in America*, 77(7), 111–28, 161.

Sparke, M., 1996. Displacing the field in fieldwork. In: N. Duncan, ed. *BodySpace: Destabilising Geographies of Gender and Sexuality*. New York: Routledge, pp. 212–33.

Spinks, P., 2014. Can a chatbot computer really deceive people into thinking it's a 13-year-old boy? *The Sydney Morning Herald*. [online] Available at: http://www.smh.com.au/it-pro/can-a-chatbot-computer-really-deceive-people-into-thinking-its-a-13yearold-boy-20140713-zt2h3.html [Accessed 14 February 2015].

Spivak, G.C., 1985. Scattered speculations on the question of value. *Diacritics*, 15(4), 73–93.

Spivak, G.C., 1999. *A Critique of Postcolonial Reason: Towards a History of the Vanishing Present*. Cambridge, MA: Harvard University Press.

Stacey, J., 2010. *The Cinematic Life of the Gene*. Durham, NC: Duke University Press.

Staeheli, L.A. and Lawson, V.A., 1994. A discussion of 'Women in the field': The politics of feminist fieldwork. *Professional Geographer*, 46(1), 96–102.

Stalbaum, B., 2006. Paradigmatic performance: Data flow and practice in the wild. *Leonardo Electronic Almanac*, 14(7–8). [online] Available at: http://leoalmanac.org/journal/vol_14/lea_v14_n07-08/bstalbaum.asp [Accessed 23 February 2015].

Stoetzler, M. and Yuval-Davis, N., 2002. Standpoint theory, situated knowledge and the situated imagination. *Feminist Theory*, 3(3), 315–33.

Stone, A.R., 1995. *The war of Desire and Technology at the Close of the Mechanical Age*. Cambridge, MA: The MIT Press.

subRosa, 2000a. *Knowing Bodies*. [multimedia installation] *Fusion! Artists in a Research Setting*, Miller Gallery, Carnegie Mellon University, Pittsburgh, PA, 22 August–29 September.

subRosa, 2000b. *Sex and Gender in the Biotech Century* [performance and website] *Digital Secrets* conference, Arizona State University, Tempe, Arizona. [online] November. Available at: http://www.cyberfeminism.net/sexgened [Accessed 23 February 2015].

subRosa, 2001a. *Refugia: Manifesto for Becoming Autonomous Zones*. [online] Available at: http://refugia.net/textspace/refugiamanifesta.html [Accessed 23 February 2015].

subRosa, 2001b. *Expo Emmagenics*. [performance and website] *Intermediale Festival: Art Happens!*, Mainz, Germany. [online] March. Available at: https://www.cmu.edu/emmagenics/home [Accessed 23 February 2015].

subRosa, 2002. Stolen rethoric: The appropriation of choice by art industries. In: M. Fernandez, F. Wilding and M.M. Wright, eds. *Domain Errors! Cyberfeminist Practices*. New York: Autonomedia, pp. 135–48.

subRosa, 2003a. *Tactical Cyberfeminism: An Art and Technology of Social Relations*. [online] Available at: http://www.artwomen.org/cyberfems/subrosa/subrosa1.htm [Accessed 23 February 2015].

subRosa, 2003b. *International Markets of flesh* [performance] *XI International Performance Art Festival: Out of Focus*, ExTeresa Arte Actual, Mexico City, 11 July.

subRosa, 2004a. In their own words. *NYFA Arts News*. [online] 10 November. http://www.nyfa.org/level3.asp?id=296&fid=6&sid=17 [Accessed 9 July 2012].

subRosa, 2004b. *Can you see us now?* [multimedia installation and website] *The Interventionists: Art in the Social Sphere*, MASS MoCA, North Adams, MA. [online] Available at: http://canuseeusnow.refugia.net [Accessed 24 February 2015].

subRosa, 2004c. *Cell Track* [multimedia installation and performance] *Bio-Difference: The Political Ecology*, Biennale of Electronic Arts Perth, Lawrence Wilson Art Gallery, Univ. of Western Australia. [online] Available at: http://refugia.net/celltrack [Accessed 24 February 2015].

subRosa, 2005a. *Selected Projects 2000–2005* [DVD] s.l.: subRosa.

subRosa, 2005b. *Epidermic! DIY Cell lab* [performance] *YOUGenics3*, Betty Rymer Gallery, Art Institute of Chicago, 18 February.

subRosa, 2009 [1999]. *SmartMom*. [website] Available at: http://smartmom.cyberfeminism.net [Accessed 23 February 2015].

subRosa, 2010 [website] Available at: www.cyberfeminism.net [Accessed 24 February 2015].

subRosa, 2011. Bodies unlimited. A decade of subRosa's art practice. *n. paradoxa*, 28, 16–25. [online] Available at: http://www.ktpress.co.uk/pdf/vol28_npara_16-25_subRosa.pdf [Accessed 24 February 2015].

Suchman, L., 2007. *Human-machine Reconfigurations*. Cambridge UK: Cambridge University Press.

Suchman, L., 2012. Configuration. In: C. Lury and N. Wakeford, eds. *Inventive Methods: The happening of the Social*. New York: Routledge, pp. 48–60.

Sui, D.Z., 2004. GIS, cartography, and the 'third culture': Geographic imaginations in the computer age. *The Professional Geographer*, 56(1), 62–72.

Takeshita, C., 2012. *The Global Biopolitics of the IUD: How Science Constructs Contraceptive Users and Women's Bodies*. Cambridge, MA: The MIT Press.

Taylor, J., 1997. The emerging geographies of virtual worlds. *Geographical Review*, 87(2), 172–92.

Teknolust, 2002 [Film] Directed by Lynn Hershman Leeson. USA: Blue Turtle, Epiphany Productions, Hotwire Productions, ZDF Productions.

Terranova, T., 2006. *Cultura network: Per una micropolitica dell'informazione*. Rome: Manifestolibri.

Terry, J. and Calvert, M., 1997. Machines/lives. In: J. Terry and M. Calvert, eds. *Processed Lives: Gender and Technology in the Everyday*. New York: Routledge, pp. 1–19.

Thacker, E., 2004. *Biomedia*. Minneapolis, MN: University of Minnesota Press.

Thacker, E., 2008. Uncommon life. In: K. Philip and B. da Costa, eds. *Tactical Biopolitics: Art, Activism and Technoscience*. Cambridge, MA: The MIT Press, pp. 309–22.

Thiele, K., 2014. Ethos of diffraction: New paradigms for a (post)humanist ethics. *Parallax*, 20(3), 202–16.

Thielmann, T., 2010. Locative media and mediated localities: An introduction to media geographies. *Aether: The Journal of Media Geography*, 5(a), 1–17.

Thrift, N., 2004. Remembering the technological unconscious by foregrounding knowledges of position. *Environment and Planning D*, 22, 175–90.

Thrift, N., 2008. *Non-representational Theory: Space/Politics/Affect*. New York: Routledge.

Ticineto Clough, P., 2003. Affect and control: Rethinking the body 'beyond sex and gender'. *Feminist Theory*, 4(3), 359–64.

Timeto, F., 2013. Redefining the city through social software: Two examples of open source locative art in Italian urban space. *First Monday*, 18(11). [online] Available at: http://firstmonday.org/ojs/index.php/fm/article/view/4952 [Accessed 25 February 2015].

Timeto, F., 2015. Locating media, performing spatiality: A nonrepresentational approach to locative media. In: R. Wilken and G. Goggin, eds. *Locative Media*. New York: Routledge, pp. 94–106.

Townsend, A., 2006. Locative-media artists in the contested-aware city. *Leonardo*, 39(4), 345–47.

Turing, A.M., 2008 [1950]. Computing machinery and intelligence. In: R. Epstein, G. Roberts and G. Beber, eds. *Parsing the Turing Test: Philosophical and Methodological issues in the Quest for the thinking Computer*. New York: Springer, pp. 23–65.

Tuters, M., 2010. *Forget Psychogeography: The Object-turn in Locative Media*. [online] Available at: http://web.mit.edu/comm-forum/mit7/papers/Tuters_DMI_MIT7.pdf [Accessed 23 February 2015].

Tuters, M., 2012. From mannerist situationism to situated media. *Convergence: The International Journal of Research into New Media Technologies*, 18(3), 267–82.

Tuters, M. and Varnelis, K., 2006. Beyond locative media: Giving shape to the Internet of Things. *Leonardo*, 39(4), 357–63.

2001: A Space Odyssey, 1968 [Film] Directed by Stanley Kubrick. USA: Metro-Goldwyn-Mayer and Stanley Kubrick Production.

United Nations, 2013. *International Migration Report*. [pdf] New York: Department of Economic and Social Affairs, Population Division. Available at: http://www.un.org/en/development/desa/population/publications/pdf/migration/migrationreport2013/Full_Document_final.pdf [Accessed 19 February 2015].

van Alphen, E.J., 2002. Imagined homelands: Re-mapping cultural identity. In: T Cresswell and G. Verstraete eds. *Mobilizing Place, Placing Mobility: The Politics of Representation in a Globalized World*. Amsterdam: Rodopi, pp. 53–70.

Velasquez, D., 1656. *Las Meninas*. [oil on canvas] (Madrid, Prado National Museum).

Verstraete, G., 2007. Women's resistance strategies in a high-tech multicultural Europe. In: K. Marciniak, A. Imre and Á. O'Healy, eds. *Transnational Feminism in Film and Media*. New York: Palgrave Macmillan, pp. 129–45.

Virilio, P., 2004. *The Paul Virilio Reader*. S. Redhead, ed., New York: Columbia University Press.

Volkart, Y., 1999. War zone. Bodies, identities and femininity in the global high-tech industry. Available at: http://www.xcult.org/volkart/pub_e/war_zones. Biemann.html [Accessed 5 April 2015].

Volkart, Y., 2000. *Technologies of Identities*. [online] Available at: http://www.obn.org/reading_room/writings/html/technologies.html [Accessed 19 February 2015].

Wajcman, J., 2004. *TechnoFeminism*, Cambridge: Polity Press.

Wallace, M., 1988. The Politics of location: Cinema/theory/literature/ethnicity/sexuality/me. *Framework*, 36, 42–55.

Weinstein, M., 2008. Captain America, Tuskegee, Belmont, and righteous guinea pigs: Considering scientific ethics through official and subaltern perspectives. *Science & Education*, 17(8–9), 961–75.

Weiser, M., 1991. The computer for the 21st century. *Scientific American*, 265, 94–104. [online] Available at: http://wiki.daimi.au.dk/pca/_files/weiser-orig.pdf [Accessed 25 February 2015].

Wertheim, M., 2000. *The Pearly Gates of Cyberspace: A History of Space from Dante to the Internet*. New York: W.W. Norton & Company.

Whatmore, S., 2002. *Hybrid Geographies: Natures, Cultures, Spaces*. London: Sage.

Whatmore, S., 2006. Materialist returns: Practising cultural geography in and for a more-than-human world. *Cultural Geographies*, 13, 600–609.

Whitehead, A.N., 1979 [1929]. *Process and Reality: An Essay in Cosmology: Corrected Edition*. D.R. Griffin and D.W. Sherburne, eds. New York: The Free Press.

Wilding, F., 2002. Vulvas with a difference. In: M. Fernandez, F. Wilding and M. Wright, eds. *Domain Errors! Cyberfeminist Practices*. New York: Autonomedia, pp. 149–60.

Wilken, R., 2005. From stabilitas loci to mobilitas loci: Networked mobility and the transformation of place. *Fibreculture*, 6. [online] Available at: http://journal.fibreculture.org/issue6/issue6_wilken.html [Accessed 19 February 2015].

Willemse, E.W., 2010. *The Phenomenon of Displacement in Contemporary Society and its Manifestations in Contemporary Visual Art*. MA Thesis, University of South Africa. [online] Available at: http://uir.unisa.ac.za/bitstream/handle/10500/4343/dissertation_willemse_e.pdf?sequence=1 [Accessed 15 February 2015].

Williams, R., 1974. *Television, Technology and Cultural Form*. London: Fontana.

Williamson, J., 1986. Woman is an island: Femininity and colonization. In: T. Modleski, ed. *Studies in Entertainment: Critical Approaches to Mass Culture*. Bloomington, IN: Indiana University Press, pp. 99–118.

Wolff, J., 1993. On the road again: Metaphors of travel in cultural criticism. *Cultural Studies*, 1(2), 224–39.

Wollen, P., 1999. Mappings: Situationists and/or conceptualists. In: M. Newman and J. Bird, eds. *Rewriting Conceptual Art*. London: Reaktion Books, pp. 27–46.

Wood, D., 2006. Map art. *Cartographic Perspectives*, 53, 5–14.

WoW, n.d [website] Available at www.womenonwaves.org [Accessed 24 February 2015].

WoW, 2012a. [website] Available at: http://www.dieselforwomen.com/reveal_ index.html [Accessed 24 February 2015].

WoW, 2012b. *A Brave New World for Female Factory Workers*. [online press release] Available at: http://www.dieselforwomen.com/revealed/about.html [Accessed 24 February 2015].

WoW, 2012c. Fashion Industry violates women's rights. [online press release] 3 February. Available at: http://www.womenonwaves.org/en/page/2573/press-release-3-2-2012-fashion-industry-violates-women-s-rights [Accessed 24 February 2015].

Writing Desire, 2000. [video essay] Ursula Biemann. Brussels: Argos Video Library.

X-Mission, 2008. [video essay] Ursula Biemann. Brussels: Argos Video Library.

Young, I.M., 1990. The ideal of community and the politics of difference. In: L. Nicholson, ed. *Feminism/Postmodernism*. New York: Routledge, pp. 300–23.

Yuval-Davis, N., 2006. Human/women's rights and feminist transversal politics. In: M. Marx Ferree and A. Mari Tripp, eds. *Global Feminisms: Transnational Women's Activism, Organizing and Human Rights*. New York: New York University Press, pp. 275–95.

Zanger, A., 2005. Women, border and camera. Israeli feminine framing of war. *Feminist Media Studies*, 5(3), 341–57.

Zeffiro, A., 2006. The Persistence of surveillance: The panoptic potential of locative media. *Wi: Journal of the Mobile Digital Commons Network*, 1. [online] Available at: http://wi.hexagram.ca/1_1_html/1_1_zeffiro.html [Accessed 23 February 2015].

Zeffiro, A., 2012. A location of one's own: A genealogy of locative media. *Convergence: The International Journal of Research into New Media Technologies*, 18(3), 249–66.

Zeffiro, A., 2015. Locative praxis: Transborder politics and activist potentials of experimental locative media. In: R. Wilken and G. Goggin, eds. *Locative Media*. New York: Routledge, pp. 66–80.

Zeitchik, S., 2013. Five days of 'Her': Building a future to feel like the present. *LA Times*. [online] 24 December. Available at: http://www.latimes.com/entertainment/movies/moviesnow/la-et-mn-spike-jonze-her-production-future-johansson-20131224-story.html#axzz2obw9eNhUandpage=1 [Accessed 15 February 2015].

Zierhofer, W., 2002. Speech acts and space(s): Language pragmatics and the discursive constitution of the social. *Environment and Planning A*, 34(8), 1355–72.

Zulato, A., 2006. Il valore del mito e della parola nella strutturazione dell'ordine interiore. *M@GM@*, 5(1). [online] Available at: http://www.analisiqualitativa.com/magma/0404/articolo_07.htm [Accessed 25 February 2015].

Index

Actor-Network Theory (ANT) 63(n), 88,
 115, 144, 159
 see also locative media, Actor-Network
  Theory, approach for; Latour,
  Actor-Network Theory
aesthetics
 as aesth-ethics 2, 16, 148, 154–5
 of cognitive mapping 30–31
 embodied digital aesthetics 151
 Haraway's and Guattari's 2, 15,
  150–56
 of location 112
 of modelling 106–7
 of objectivity 101, 113
 and the technological 150, 154–5
 of technospaces 149(n)
 videogame 71
 see also diffraction, aesth-ethics of;
  epistemology, and aesthetics;
  Guattari, aesthetic paradigm;
  Guattari, machines, aesthetic;
  maps, and the visual arts (aesthetic
  use of)
affect 13, 30, 43, 61, 90, 129, 133, 135,
 150–52, 154
 affective labour 123
 'affectuality' 151
 and the body 30, 41, 153–4
 of/and code 43–4, 154
 in non-representational theory 61, 63
 Spinoza's concept of 154
 and the virtual 150(n)
 see also Guattari, affect;
  representation/al, and affect;
  Schwartz, affective position of the
  artist
alterity/otherness 23, 31, 36, 54, 65, 151,
 160–62
 absolute vs. relational 53, 65
 as becoming with 144

and identity 16, 25, 35, 65, 70, 144
 relations of 8, 15–16, 37–8, 49, 140,
  144, 148, 151, 154–5
 see also da Costa, *Dying for the Other*;
  Haraway, significant others; mirror,
  and otherness; place, of the other
animal 122, 144
 genomes 128
 human–animal 15,61, 141, 144–5, 147
 laboratory mice 141–3, 145, 147
 rights 140–41, 145
 see also da Costa, *Pigeon Blog*;
  Haraway, companion (species);
  Haraway, OncoMice™; Haraway,
  significant others
apparatus 130(n)
 material-semiotic 68, 71, 133, 147,
  159
 of observation 2–3, 11, 68
Art is Open Source (AOS, Salvatore
 Iaconesi+Oriana Persico) 14,
 134–9, 144, 147–8
 La Cura. The Cure 14–15, 137–9
 Human Ecosystems 135
 see also medicine, cure
art
 locative 15, 112–20, 140
 performance art 14, 125–8
 -science 14, 122, 126–30, 140–48
 spatial imagination 103–4, 113
 and system theory 151(n)
 visual arts 14, 103, 114, 123
 see also bio, art; cartography, and the
  visual arts; feminism/feminist, art;
  locative media, artistic genealogy
  of; maps, and the visual arts
  (aesthetic use of); medicine, art of;
  recombinant, art; representation,
  performative/performed, and

performance art; surveillance, art;
  vision/visuality, arts
artificial
    Artificial Intelligence (AI) 43, 54–5
    feminised 43
    Artificial Life (AL) 49, 54–5, 132
    vs. authentic 114
    Intelligent Operating System (OS) 37,
      42, 45
    perspectiva artificialis 102
    see also knowledge, Artificial
      Intelligence (AI)
assemblage 15, 58, 61, 76, 115, 129, 136,
      137, 148, 150–51, 154, 158
    machinic 37, 150–51, 154, 161
    see also configuration/s, as self-
      reflexive tool and method
      assemblage; Guattari, machinic,
      assembalges; sociotechnical,
      assemblages
autopoiesis 21, 132, 150–51, 154, 162
    autopoietic body 131–2, 134
    autopoietic system 8, 13, 15, 49, 61,
      156, 161

Barad, Karen
    agential realism 60, 69
    agential separability 12
    on Bohr's 'modified' two-slit
      experiment 68
    and Haraway's diffraction 5, 12, 60,
      67–9, 100, 149, 155, 159
    intra-action 11, 37, 59–60, 69, 71, 100,
      104, 115, 130, 149, 152
    matter/ing 3, 11–12, 5, 59, 62, 67, 69,
      79
    on mediation vs. Haraway 11, 58–60
    nonrelativist realist position 59(n)
    on representation 11, 26, 58–60, 67–8,
      78–9
    on Schrödinger's cat paradox
      experiment 3
    together-apart 8, 42
    'Together Apart (as One
      Movement)' 36–49
becoming 5, 8, 15, 22, 42, 63, 65(n), 69,
      149, 151
    philosophies of 62

-with 7, 144, 148
    see also alterity, as becoming with;
      difference, becoming-with;
      differential, becoming; space/
      spatial, becoming space
Biemann, Ursula 9, 92–101
    The Black Sea Files 93
    Europlex 98
    Performing the Border 94, 98, 99–100,
      123
    Remote Sensing 94–7, 100
    video essay 9, 92–101, 123
    Writing Desire 94, 97–8
    X-Mission 93
    see also cartography, Biemann's
      feminist; sex/sexual, economies in
      Biemann's work; video essay
bio
    art 126
    biological/technological 38, 44, 46
    biology 129, 133
      biological philosophy 62
        contestational biology 140
    and geo 57
    politics 118, 127, 133
      and necropolitics 143
      technobiopolitics 121
    power 121, 133
      technobiopower 13,14, 121, 123,
        128–9, 158
    sciences 62
    social order 132
    technobiobodies 13, 126, 133, 139
    technologies 14, 121–3
    see also Foucault, biopolitics;
      Foucault, biopower; Guattari,
      machines, biological and
      non-biological; Haraway,
      technoscience, technobiopower;
      Nold, *Bio Mapping*; performative/
      performativity, redefinition of
      technobiobodies
body/embodiment 13–15, 28, 30, 41, 46,
      56, 62, 68, 82, 94, 113, 121–48,
      151(n), 156, 158
    in control society 132–4
    and diffraction 68
    female/women's 43, 52, 56, 97–9, 125

and the feminist politics of location
   52–6
and mediation 44, 56, 58, 60, 71
national body 98
performativity 15, 59, 125–6, 132, 134
and place 17–19
Ruby's in *Teknolust* 44
Samantha's in *Her* 46–9
of theory 61
turbulent 132
of vision 60, 79, 161
See also aesthetics, embodied digital
   aesthetics; affect, and the body;
   autopoiesis, autopoietic body;
   bio, technobiobodies; borders/
   boundaries, bodily; connection/s,
   bodily; cyber, space, cyborg
   bodies in technospaces; Deleuze,
   and Guattari, body; environment,
   bodily environmentalism; image,
   bodily; information, and the body;
   medicine, body in; network/ed,
   body; objectivity, embodiment;
   sex/sexual, and the body; subRosa,
   Knowing Bodies; system, body as
Bolter, Jay and Grusin, Richard
   on non-places 35
   remediation 7, 34–5
   see also representation/al, and
      remediation
borders/boundaries
   articulation/mediation/reworking 3, 5,
      12, 15, 16, 36, 38, 60–61, 69, 78,
      86, 87, 89, 100–101, 123, 145, 155,
      158–9, 160, 161
   bodily 15, 121, 133–4, 135, 147
   dissolution 6, 29, 30, 61–2
   mobility across 10, 13, 24, 30, 92, 93,
      94–9, 114, 115–20, 135
   national/transnational 52, 72, 83, 95,
      98–9, 106, 115
      USA-Mexico border 10, 99–100,
         117, 120, 123, 140
   performativity 11, 98–9, 119
   porous 82, 129, 133–4
   see also Biemann, *Performing the
      Border*; Electronic Disturbance
      Theater (EDT), *The Transborder*

Immigrant Tool; Haraway,
   boundary breakdowns; Oliver,
   Border Bumping; place, as
   bounded/closed dimension;
   surveillance, at the border
Bridle, James
   Dronestagram 106
   Robot Flâneur 114
Butler, Judith
   on censorship 77–8
   performance/performativity 59, 62,
      147, 150

cartography
   Biemann's feminist 9, 92, 94, 100
   cartographic anxiety 107
   critical 107, 109
   navigational 9, 157
   representational 18, 92, 107, 113, 116
   schizoanalytic 30(n), 149
   situational 6
   and the visual arts 101–3, 113–14
   see also Haraway, multidimensional
      cartography/maps
Casey, Edward
   critique of Foucault 5, 19–20
   fate of place 5, 17–19
   see also site/s, Casey's definition
chatbot 39–40, 44
computer
   CT (Computerised Tomography) 135
   graphics 116
   human/computer intelligence 8,
      40–41, 49
   and the informatic environment 42,
      132, 148
   and representationalism 24–6
   revolution 51
   see also recombinant, computing;
      screen, computer (in *Her*);
      Turing, 'Computing Machinery
      and Intelligence'; Turing, test;
      ubiquitous, computing/information
configuration/s 3, 5, 8, 11, 15, 16, 59,
   69, 79, 80, 83, 87, 105, 106, 107,
   117(n), 118, 130, 152, 155, 159,
   160

as self-reflexive tool and method
    assemblage 155
see also figuration/s, as con-
    figurations; sociotechnical,
    configurations; knowledge,
    configuration/practice/production/
    process
connection/s
    bodily 46
    connectivity 32, 33, 69, 85, 123, 149,
        161
        vs. engagement 149
    and disconnection/separation 32,
        36–49, 82, 100, 114, 137, 160–61
    the earth-wide network of 58
    machinic 8, 49
    making/tracing/performing/weaving 9,
        13, 15, 123, 30, 36, 39, 65, 69, 73,
        86, 92, 94, 122, 124, 137, 139, 143,
        145, 159, 160
    multiple/multiplication of 7, 14, 15,
        16, 72, 87, 122
    n-connections 90
    partial 64, 154–5
    in technospaces 7, 16, 154, 159
control
    resistance 13, 15, 96, 106, 117, 129,
        133, 140
    society 132–3, 149, 150
    spatial 20, 90–91, 93, 131–6
    technologies of 96, 99, 106, 116–17,
        121–2, 126–7, 132–6, 140, 150,
        158
    see also Deleuze, 'Postscript of the
        Societies of Control'; network/ed,
        control/power
creative
    capacity/force/potential 1, 2, 11,
        15–16, 30(n), 58, 106, 116,
        150–51, 153, 158, 160–62
    science 126, 128, 130, 145
    see also imagination, creative;
        performative/performativity,
        creativity
Critical Art Ensemble (CAE) 126, 127,
    139–40
    see also bio-, biology, contestational
        biology; da Costa, and CAE

cyber
    brides 96
    cybernetics/cybernetic theory 13, 41,
        49, 132
        loops 8, 21
        third wave cybernetics 49, 132
    feminist 94, 122–30
    space 24–5, 35, 87, 97, 104, 106–7,
        148
        cyborg bodies in technospaces 94,
            121–3
        'Cyborgs and Space' 52
    see also Haraway, cyborg; Hayles,
        third wave cybernetics
da Costa, Beatriz
    and CAE 139–40
    Dying for the Other 14–15, 139,
        141–8
        Haraway on 144, 147
    Pigeon Blog 140–41, 144–5
    see also diffraction, in *Dying for the
        Other*

Deleuze, Gilles 26(n), 62
    figural 65(n)
    and Guattari, Félix,
        body 46
        machinic phylum 150(n)
        multiplicity 22
        nomadology 18
        striated and smooth space 22, 129
    on Leibniz 18
    'Postscript of the Societies of Control'
        132
    see also body/embodiment, in control
        society; control, society; medicine,
        in control society
difference
    becoming-with 15, 42, 149
    emerging 38, 63, 148
    epistemic 10, 55(n)
    evaluation of 128
    in feminist scholarship 57
    and the figural 65(n)
    of the inappropriate/d other 101
    and location 10, 53, 56, 57, 66, 160
    making 16, 67–8, 80, 82, 144, 155

and the methodology of diffraction 5,
12, 67–8, 70–71, 78, 94, 116, 144,
155
power-related 85, 96
class 31(n)
sex 91
relations/relating 20, 41, 66, 148, 161
and repetition 146
visual/visual construction of 94, 100,
103, 157
see also power, of differentiation;
sex/sexual, social structuration of
sexual difference; signs, different
order of; signs, different practice
of; space/spatial, differences/
heterogeneity/multiplicity
differential
articulation 59, 78
becoming 8, 69, 79
condition 18
consciousness 120
forces 18, 151(n)
imaging 151
location/position 88, 104
relations 18, 120, 154
diffraction
aesth-ethics of 2, 153, 155–6
'The Aesth-ethics of Diffraction'
148–56
of criticism 130
'Diffracting Technoscience' 121–56
diffractive metaphor 153
diffractive representations 11, 16, 67,
139, 152, 160
in *Dying for the Other* 144–7
in media and visual studies 12, 70–73
methodology of 1, 5, 12–13, 67–8,
70–71, 73, 75, 78, 80, 100, 149,
159–60
'Mattering Light: A Diffractive
Methodology' 65–70
in *Mirror Image* 13, 73, 75, 78, 80
physical phenomenon of 2–4, 67–9,
159
and the politics of location 1, 152, 155
see also Barad, and Haraway's
diffraction; body/embodiment,
and diffraction; difference, and

the methodology of diffraction;
figuration/s, and diffraction;
Haraway, diffraction; Haraway,
location and diffraction; maps,
of diffraction; reflection,
representational vs. diffraction;
screen, and diffraction; video essay,
and diffraction
displacement
geographical 9, 13, 19, 74, 76, 93, 94
identity/sameness/subject's position
15, 54, 68, 69, 100, 116, 155, 160
images and words 78, 81
mirror in *Mirror Image* 73, 76, 83
in representation 6, 12, 116, 127, 134,
152, 155, 159, 160
distance
end of critical distance 6, 30, 33–4,
105
representational/visual 6–7, 12, 46, 59,
67, 78, 96, 97, 101, 102, 105, 112,
130, 146, 152, 156, 157, 160–61
spatial 22, 25(n), 30, 33, 87, 105, 132,
152
see also vision-visual/ity, distance/ing

ecosystem/ic 14, 130, 134–7, 139, 148,
153, 159, 161
approach 139, 148
see also Art is Open Source, *Human
Ecosystems*; medicine, cure, 'An
Ecosystemic Cure'; network/ed,
ecosystemic
Elahi, Hasan M.
Tracking Transience 116–17
Electronic Disturbance Theater (EDT)
b.a.n.g. lab 118, 119
Electronic Civil Disobedience (ECD)
117, 120
The Transborder Immigrant Tool 10,
117–20, 121
entanglement 11–13, 58–9, 62, 68, 71,
78, 86, 110, 133, 146, 152, 155–6,
158–9, 161
in quantum physics 2–5
see also measurement, entanglement of
entropy 131–2, 149
see also information, and entropy

environment
    bodily environmentalism 133
    media 25, 42, 86–90, 105, 110, 112,
        115, 158
        informational/informatic 14, 20,
            111, 132, 149(n), 161
        multimodal 126
        smart 36–7, 105
    natural 58, 140–41
    subject-environment 6, 25, 30, 36–7,
        45–6, 48–9, 69
    system-environment 61, 64, 132
    urban 115, 137
    see also computer, and the informatic
        environment; Jeremijenko, *The
        Environmental Health Clinic*;
        sociotechnical, environments;
        technology/ies, background/
        environment
epistemology
    and aesthetics 153–4, 161
    of complexity 131
    constructivist 54
    ecological 23, 58
    epistemological modesty 56
    and ethics 5, 53, 57, 58, 149
    feminist 2, 54, 58, 63
    and ontology 3, 5, 60, 68, 149, 160
    postmodern 26
    representational 27, 59, 115
    situated 58, 128, 138
    standpoint 2, 10, 54–5, 57, 59(n)
    see also difference, epistemic;
        Harding, standpoint epistemology;
        objectivity, standpoint
        epistemology; relativism, in
        standpoint epistemology
ethics/ethical 5, 14, 53, 56, 57, 58, 63,
    79, 100, 118, 136, 137, 140, 144,
    149–52
    see also aesthetics, as aesth-ethics;
        epistemology, and ethics;
        representation, politics of, and
        ethics

feminism/feminist
    activism 71, 123, 125
    art 14, 122–30

art history 91
    performance 125–7
constructivism 10
material 57–8, 71
networking 122–3
and non-representational theory 62–3
objectivity 56
postcolonial transcultural 57
science (and technology) studies 56,
    58, 128–9
    technoscience 57
socio-technical perspective 38
spatialisation of 51, 53
theory/theories/theorists/thought 2, 10,
    51–3, 66, 120
    Third Wave 51
    White Western 52–3
transversal politics 155
see also body/embodiment, and the
    feminist politics of location;
    cartography, Biemann's feminist;
    cyber, feminist; difference, in
    feminist scholarship; epistemology,
    feminist; geography, feminist;
    Harding, *The Science Question
    in Feminism*; location, politics
    of, feminist; place, feminist
    conceptions
field
    of forces 5, 131
    implication in the 9, 55–6, 101, 104
    material-semiotic 56
    of possibilities 64, 132, 160
    of relations 36, 38, 133, 137, 157
    representational/visual 12, 63, 69, 70,
        77, 96, 101, 103–4, 131, 157, 159
    spatial 18–19, 21, 56, 75, 103–4, 108
figuration/s 12, 65–7, 69, 70, 100, 160
    as con-figurations 65
    and diffraction 12, 16, 67, 69, 159
    of hybrid subjectivity 94
    vs. metaphors 65
    as topoi and trópoi 12, 66, 152
    see also Haraway, figurations;
        Haraway, location, and figurations;
        maps, figurations as; performative/
        performativity, figurations; power,
        of figurations; time, of figurations

flâneur/flânerie 91, 114–15
  see also Bridle, *Robot Flâneur*
Foucault, Michel
  biopolitics 133(n)
  biopower,121
  The Birth of the Clinic 130–31
  geography 20
    critique of Foucault's geography in
      Harvey 27(n)
  heterotopia 1, 5–7, 19–20, 33, 53, 110
    as mirrors 20, 25, 28–9, 33, 36, 38,
      78, 161
  'Of Other Spaces' 19–20
  panoptical model 105, 131
  vs. phenomenology 20
  power relations in space 20, 33, 131
  spatialisations of medical knowledge
    13, 130–31
  temporality of space 20, 29, 33
  see also Casey, critique of Foucault;
    site/s, and countersite in Foucault

generativity
  spatial 1, 7, 21, 76, 89–90
  of representation and visual practices
    4, 7, 12, 61, 67, 69, 71, 79, 106,
    160
geography
  feminist 6–7, 57, 103
  geomedia 2, 9, 110
  human 51
  metaphors 85
  neogeography 110(n)
  and non-representational theory 57–8,
    62–3
  as not-merely spatial discipline 7, 20,
    27
  ocularcentrism 101–2
  of survival 100
  see also displacement, geographical;
    Foucault, geography; geospatial,
    technologies; Gregory,
    Geographical Imaginations;
    Mitchell, T.; information,
    geolocalised; place, and geomedia/
    locative media; Situationism,
    psychogeography

global
  capitalism/economy 99, 123
  imagination/description of
    globalisation 32, 85, 87
  and local 53, 90–91, 129
  networks/flows/routes 9, 31, 88, 96,
    121, 123
  of feminised labour 95
  sex trade and industry 95
  positioning 115–16
    systems (GPS) 9, 112, 119
  village 19
  see also location, and global networks
Graham, Stephen
  perspectives on space and information
    technologies 87–8
Gregory, Derek
  Geographical Imaginations 101–7
  on the world as exhibition 101–5
  on Harvey 85
  on Jameson 31
grid
  of hyper-geo-mapping-power 10, 118
  model 18, 85, 96, 104, 113, 116
  as graticule 104(n)
Guattari, Félix
  aesthetic paradigm 149(n), 151, 154,
    159
  affect 150–51, 154
  ecosophical logic 153
  Integrated World Capitalism 149(n)
  machines
    aesthetic 15, 150–51
    biological and non-biological 148
    existential 49
    schizonalytic 8
    semiotic 21
  machinic 2(n)
    assemblages 37, 151, 154
    interconnectedness 149
    vs. mechanic 150
  meta-modelisation 21, 22(n), 150
  see also aesthetics, Haraway's
    and Guattari's; cartography,
    schizoanalytic; Deleuze, and
    Guattari; machine, machinic
Haraway, Donna
  boundary breakdowns 61–2

companion (species) 16, 144, 145,
    147, 148, 149, 153, 161
    respect 148, 162
cyborg 133
diffraction 1, 2, 5, 11,12–13, 60, 65,
    67–70, 72, 75, 78, 94,100, 106,
    116, 139, 149, 152, 155–6, 159
figurations 12, 16, 65–7, 69, 70, 152,
    159–60
inappropriate/d other 65, 100–101
integrated circuit 13, 122–4, 149, 154
location 11, 56–7, 60
    and diffraction 1, 69, 152, 155,
        159–60
    and figurations 12, 66, 152
Modest Witness 67
    modest witness 145
mediation 11, 56, 58, 60, 61, 62, 66,
    72, 75–6, 154, 159
multidimensional cartography/maps
    55, 63, 152–3
objectivity 56–7, 63, 79
OncoMice™ 147
optics 5, 60–61, 67–8, 79, 149, 155,
    159
'The Promises of Monsters' 67, 72
representation 10, 11–12, 61, 63, 67,
    69, 152, 156, 159
significant others 14, 140, 143–5
    'Significant Others' 139–48
situated knowledges 2, 5, 11, 53, 56–8,
    59(n), 60, 63, 128–9, 159
    'Situated Knowledges' 56, 67
technoscience 13, 56–7, 65, 66, 68, 79,
    106, 121, 129, 147
    technobiopower 121, 129
theory of vision,11, 12, 60, 61, 63, 67,
    69, 79, 106, 149, 152–3
virtual 151(n)
see also aesthetics, Haraway's and
    Guattari's; Barad, and Haraway's
    diffraction; Barad, on mediation
    vs. Haraway; connection/s, the
    earth-wide network of; da Costa,
    Dying for the Other, Haraway on;
    difference, of the inappropriate/d
    other; diffraction, methodology of;
    figuration/s

Harding, Sandra
    The Science Question in Feminism 56
    standpoint epistemology 2, 10, 54–5,
        57, 59(n), 60
    strong objectivity 10, 55
    see also epistemology, standpoint
Harvey, David
    time-space compression 29, 85, 104
    See also Foucault, geography, critique
        of Foucault's geography in Harvey;
        Gregory, on Harvey
Hayles, Katherine
    constrained constructivism 11, 63–5,
        69
    inscription and incorporation 8, 48
    intermediation 71
    interpretation of the Turing test 8, 41
    third wave cybernetics 49
    see also real/reality, in constrained
        constructivism
heterotopia/heterotopic 1, 5–7, 19–20, 23,
    25, 28, 33, 36, 39, 53, 110, 161
    and heterochronies 33
    non-places as heterotopias 36
    technospaces as 6
    see also Foucault, heterotopia; place,
        and the heterotopic mirror; real/
        reality, and heterotopias; space,
        heterotopic

image
    bodily 133–5
    Digital Imaging and Communications
        in Medicine (DICOM) 135, 137(n)
    Magnetic Resonance Imaging (MRI)
        135
    mirror(ed) 25, 28, 83, 106–7
    physics experiments on imaging 3–4
    the place of the 63, 81, 83
    satellite 96, 106, 114, 117
    truth of the 72, 83, 94
    see also displacement, images and
        words; Schwartz, *Mirror Image*;
        space/spatial, energy-space
        imaging
imagination
    creative 15, 153–4
    situated 155–6, 159

spatial 2, 8, 22–3, 25, 26(n), 32, 51,
85–6, 103, 108, 120
see also art, spatial imagination;
differential, imaging; global,
imagination/description
of globalisation; Gregory,
Geographical Imaginations;
place, imagined; representation/al,
imagination/imaginary
information 34
accessible 136–7, 140
and the body 2, 13, 15, 48–9, 97, 118,
131–4, 145
circulation/movements/flows/networks
13–15, 72, 88, 89, 97, 110, 113,
131, 132, 133, 134, 136, 139, 144,
158
and entropy 132
geolocalised 110–11
Geographical Information Systems
(GIS) 97, 105, 111
Volunteered Geographical
Information (VGI) 111
as in-formation 8, 88, 136, 139
informational space 132
territory 110
and materiality/matter 3–4, 8, 21, 31,
56, 88, 89, 110, 134, 160, 161
modulation 132–3
performativity 14, 137
situation 86–7
society 88, 89,110
technologies/communication
technologies (ICTs) 8, 9, 24, 29,
38, 85, 87–8, 96–7, 99, 104, 108,
113
see also environment, media,
informational/informatic; Graham,
perspectives on space and
information technologies; machine,
informatic; mobility, migratory/
transnational and ICTs; network/ed,
information; place, and information
flows/technologies; situation,
of information; transnational,
economies and ICTs; ubiquitous,
computing/information

interface
graphical user interface (GUI) 34
limitless 151
maps/locative media as 108–10, 116
mediation 1, 6, 7, 8, 20–21, 34, 36–7,
87, 151, 161
baroque 37, 90
in *Her* 8, 37–49
as intraface 37
in *Teknolust* 8, 44
see also mirror, as interface; non-
representational, interface;
performative/performativity,
interface; topological/topology,
interface; screen, as interface;
ubiquitous, interfaces
intra-action 11, 37, 48, 59–60, 69, 71, 100,
104, 115, 130, 149, 152, 158, 159
see also Barad, intra-action

Jameson, Fredric
cognitive mapping 1, 6–7, 30–31,
32(n)
representational tropes 31
theory of postmodern space 29–31
critiques of 31–4
see also aesthetics, of cognitive
mapping; distance, end of critical
distance; Gregory, on Jameson
Jeremijenko, Natalie
The Environmental Health Clinic
136–7
Jonze, Spike
Her 8, 37–49
see also body/embodiment, Samantha's
in *Her*; interface, mediation, in
Her; real/reality, Samantha in *Her*;
screen, computer (in *Her*); sex/
sexual, relationship in *Her*

knowledge
archaeology 20
Artificial Intelligence (AI) 43
configuration/practice/production/
process 10–11, 14, 23, 55–6, 58–9,
61–2, 69, 75, 76, 79, 80, 121, 126,
130–31, 139, 144, 145, 153–6

dissemination/sharing 14, 128–9, 134, 139
medical 13, 130–31, 133
politics of 61, 66
and responsibility 57, 128
scientific 10, 55–6, 58, 59, 126, 130
and space 104, 107, 111, 115, 130–33
subjects/objects relation 11, 14, 36, 54–5, 59, 76, 78, 100, 130, 139, 161
and transparency 34, 36
see also Foucault, spatialisations of medical knowledge; location, of knowledge; performative/ performativity, of knowledge practices; situated knowledge/s

laboratory
clinic 131, 136
open 14, 126, 145
'The Open Lab' 121–30
scientific 15, 141–2, 145, 147
see also animal, laboratory mice; Electronic Disturbance Theater, b.a.n.g. lab; Latour, World Wide Lab; subRosa, *Epidermic! DYI Cell Lab*
Latour, Bruno 76, 90, 105
Actor-Network Theory 63, 80, 115
matters of fact, matters of concern 129–30
mediators vs. intermediaries 48
second empiricism 129
maps 27–8, 108–9
World Wide Lab 134
see also Actor-Network Theory
Leeson, Lynn Hershman
Teknolust 8, 43–4
see also interface, mediation, in *Teknolust*; self-replicating automatons, in *Teknolust*
light
mattering 12, 67–8
metaphysics of 66–7, 152
nature of 3–4, 68, 159
see also diffraction, methodology of, 'Mattering Light: A Diffractive Methodology'

location
ecological consideration of 11
and global networks 85–91
locative networks 111, 113, 115, 117(n), 119, 131
network locality 9, 85, 87, 90
knowledge 2, 10, 122, 126, 129
of the subject's/speaker's voice 9, 13, 93–4, 147
and mobility 2, 9, 53, 72, 89–90, 92–4, 96–8, 103, 108, 111, 112
'Location, Mobility, Perspectives' 85–120
politics of,
feminist 1, 2, 10–11, 51–7, 60, 66, 91, 123, 144–5, 159–60
postcolonial 53–4, 57
in technospaces 9, 87, 89–90, 92, 129
see also aesthetics, of location; art, locative; body/embodiment, and the feminist politics of location; difference, and location; differential, location/position; diffraction, and the politics of; global, and local; Haraway, location; locative media; representation/al, relocation of; Rich, 'Notes Towards a Politics of Location'; technology/ies, locative
locative media 1, 2, 9
Actor-Network Theory approach for 115
artistic genealogy of 15, 112–15
dis-locative 116, 117(n), 130, 131
dislocating 94
as 'embedded media' 113, 116
'A Politics of Location for Locative Media' 110–20
technical definition 111–12
see also art, locative; interface, maps, locative media as; place, and geomedia/locative media; Situationism, revival in locative media
Lynes, Krista Geneviève
prismatic media 70

machine
    human/machine relationship 1, 8,
        36–8, 40–41, 46, 49, 55, 61, 154,
        158, 161
    informatic 132
    machinic 8, 16, 37–8, 49, 98
    Third Machine Age 29
    see also assemblage, machinic;
        connection/s, machinic; Deleuze,
        and Guattari, machinic phylum;
        Guattari, machines; Guattari,
        machinic; technology/ies, machines
Magritte, René
    La Reproduction Interdite 28
    Les Liaisons Dangereuses 28
maps
    of diffraction 68
    of diseases 130
    figurations as 12, 66
    non-representational 8, 27–8
        navigational 8, 9, 108–9
    performative mapping practices 15,
        35, 107–8, 110, 120
        location-based applications/
            locative media 15, 109–13,
            116, 119
            experiential mapping 112
        mapping technologies 14
    representational 120
    and territories/spaces 8, 24, 29, 35,
        107–11, 113–14, 119, 137, 148
        map spaces 109
    and the visual arts (aesthetic use of)
        15, 101, 109, 113–14, 123–4, 128
    see also aesthetics, of cognitive
        mapping; grid, of hyper-geo-
        mapping-power; Haraway,
        multidimensional cartography/
        maps; interface, maps, locative
        media as; Jameson, cognitive
        mapping; Latour, maps;
        mediation, maps; Nold, *Bio
        Mapping*; space/s, map spaces;
        ubiquitous, maps
Massey, Doreen
    spatial theory 6,7, 20, 22, 23, 25–6,
        32–3, 57, 86–7, 91
    coevalness 6, 7, 23, 32–3, 57

matter/materiality 30, 36, 42, 52, 58, 62–3,
        78–9, 151(n)
    dual behaviour 2, 68
    mattering 3, 11–12, 59, 62, 66–7, 69,
        88, 152–3, 160
    mind/matter 26–7
    see also Barad, matter/ing; feminism/
        feminist, material; field, material-
        semiotic; information, and
        materiality/matter; light, mattering;
        network/ed, digital and material;
        performative/performativity,
        matter; real/reality, material/realm
        of things/substantial/world; space/
        spatial, abstract-symbolic/material
measurement
    entanglement of 3–4, 12
    spatial 5, 19, 21–2, 33, 34, 66, 87, 90,
        132
    see also system, measuring
media 19, 21, 30, 71, 72, 101, 104, 108,
        119
    convergence 112
    digital 1, 85, 87, 105
    electronic 86
    enactive 92, 104–5
    grassroots vs. official 70
    high and low-tech 123
    medium-medium relations in non-
        representational theory 62
    new 91, 122, 145
    performative 110
    tactical 118
    theory/studies 7, 12, 110(n), 145, 154
    visual and visual technologies 104–5,
        115
    see also diffraction, in media and
        visual studies; environment, media;
        geography, geomedia; locative
        media; Lynes, prismatic media;
        mediation, media; Mitchell, W. J.
        T.; space/spatial, spatial turn of
        media studies
mediation
    maps 27, 107, 109
    media 70–73, 101, 104, 118–19, 158
    mirror 25, 27–9, 36–7, 75, 161
    sociotechnical 20, 36, 37, 89, 150

in technospaces 1, 2, 6, 7, 8, 9, 15, 36,
    87, 88, 115, 158, 161
of vision 11, 12, 60, 105, 72, 79, 158
see also Barad, on mediation vs.
    Haraway; body/embodiment,
    and mediation; Bolter and
    Grusin, remediation; borders/
    boundaries, articulation/mediation/
    networking; Haraway, mediation;
    Hayles, intermediation; interface,
    mediation; Latour, mediators
    vs. intermediaries; non-human,
    and human mediations/relations;
    performative/performativity,
    mediations/mediated practices;
    real/reality, and mediation;
    sociotechnical, mediations
medicine
    antipsychiatric movement 135
    art of 153
    body in 14–15, 128, 130, 132–5, 147
    in control society 132–3
    cure 14, 130, 147, 148
        La Cura. The Cure 137–9
        'An Ecosystemic Cure' 130–39
        politics of who and politics of what
            15, 144–5
    disease 14, 15, 130–36, 143–5, 147
    historicity of 130
    medical abortion 123–4
    medical gaze 130–31
    patient 14, 81, 82, 130, 132, 135–6,
        138–9, 143, 144–7
        impatient 136
    urban acupuncture 137
    see also Art is Open Source, *La Cura.
        The Cure*; da Costa, *Dying for the
        Other*; Foucault, *The Birth of the
        Clinic*; Foucault, spatialisations of
        medical knowledge; image, Digital
        Imaging and Communications
        in Medicine; image, Magnetic
        Resonance Imaging; Jeremijenko,
        *The Environmental Health
        Clinic*; knowledge, medical;
        laboratory, clinic; maps, of disease;
        representation/al, of disease;

surveillance, medicine; technology/
    ies, reproductive; truth/s, of
    disease; Women on Waves,
    Misoprostol campaign
mirror
    as interface 13, 25, 36, 39, 75, 80, 83
    and otherness 25, 27, 35–6, 161
    see also displacement, mirror in *Mirror
        Image;* Foucault, heterotopia,
        as mirrors; image, mirror(ed);
        Magritte, *La Reproduction
        Interdite*; mediation, mirror;
        place, and the heterotopic mirror;
        reflection, mirror; Schwartz,
        Mirror Image
Mitchell, Timothy
    'Geography and the World-as-
        Exhibition' 101–3, 105
    see also Gregory, *Geographical
        Imaginations*, on the world as
        exhibition
Mitchell, W.J.T.
    addressing media 89
mobility
    freedom of 85, 93
    migratory/transnational and ICTs
        9–10, 92–3, 92–9, 117–20, 123
    in technospaces 86, 88, 92, 108, 111
    women's 91, 92, 95–9, 124, 128
    see also borders/boundaries, mobility
        across; location, and mobility;
        network/ed, mobility; technology/
        ies, mobile
modernity 52, 87, 107
    Second 93
    Western 26, 29(n), 32(n), 62, 91,
        103–4, 157
network/ed
    body 44, 62, 122, 134
    control/power 13, 106, 120, 121, 123,
        129, 133–4
    digital 20, 94
    and material 94, 108
    ecological 16, 153
    ecosystemic 14
    information 9, 136, 139, 158

and communication 13, 98, 131,
133
mobility 92
place 6, 88, 90
of practices 16, 146, 153
social/society 29, 70, 86, 135
sociology 87
sociospatial 87, 134
sociotechnical 9, 88–9
space 63
see also Actor-Network Theory;
connection/s, the earth-wide
network of; feminism/feminist,
networking; global, networks/
flows/routes; information,
circulation/movement/flows/
networks; location, and global
networks; ubiquitous, networks

Nold, Christian
Bio Mapping 118(n)
non-human
and human mediations/relations 11,
37, 50(n), 88, 115, 140, 147, 150,
151, 161
non-representational
approach/theory 1, 8, 11, 13, 15, 27,
59(n), 61, 62–3, 69, 89, 149, 159
interface 39
perception 64
see also affect, in non-representational
theory; feminism/feminist,
and non-representational
theory; geography, and non-
representational theory; maps,
non-representational; media,
medium-medium relations in
non-representational theory;
representation/al, in non-
representational theory; Thrift,
non-representational theory
North American Free Trade Agreement
(NAFTA) 99, 118, 123

objectivity
embodiment 131
performed 12, 59

scientific notion of 55, 57, 59, 68, 145,
149
standpoint epistemology 55–6
see also aesthetics, of objectivity;
feminism/feminist, objectivity;
Haraway, objectivity; Harding,
strong objectivity
observation
distant/detached/external 23, 46, 59,
78, 96, 101–2, 106, 109
observer's position 12, 41, 69, 78,
101–2, 103, 109, 111
observing-observed coimplication 2–3,
6, 11–12, 26–7, 34, 37, 40–41, 59,
64, 68–9, 71, 78–9, 90, 100, 103–4,
142, 153, 159
second order 151–2(n)
and self-observation 105
space of 9, 27
see also apparatus, of observation;
performativity, of observation
Oliver, Julian
Border Bumping 117(n)
Oppenheimer, Joshua
The Act of Killing 81
The Look of Silence 80–82
optics
binary 100
geometrical 68, 149, 159
oligoptism 105
physical 3–4, 68, 159
production of positioning 79
sciences 104
spatial 107
see also Foucault, panoptical model;
Haraway, optics
orientalism 7, 31, 101

Parker, Cornelia
Unsettled 76
performative/performativity
approach 14, 15, 40(n), 71, 89, 105,
107, 134, 137, 139, 159
creativity 15
vs. expression 29, 37, 150, 153
figurations 65–6, 69
idiom 11, 62, 69

interface 37
of knowledge practices 58
matter 3, 59
mediations/mediated practices 7,
     60–61, 66, 71, 158
of observation 100
redefinition of technobiobodies 13
space 1, 2, 6, 9, 23, 26(n), 29, 54,
     107–9, 110, 119, 120, 152
see also art, performance art; Biemann,
     Performing the Border; body/
     embodiment, performativity;
     borders/boundaries, performativity;
     Butler, performance/performativity;
     connection/s, making/tracing/
     performing/weaving; feminism/
     feminist, art, performance;
     information, performativity; maps,
     performative mapping practices;
     media, performative; objectivity,
     performed; representation/al,
     performative/performed; subRosa,
     performances; technology/ies,
     performative
place
   as bounded/closed dimension 9, 17,
        81, 91–2
   commonplace/s 12, 152, 156, 160
   extroverted sense of 86
   feminist conceptions 51–2, 54, 75, 91
   gendered 91, 103
   and geomedia/locative media 110–13,
        116
   and the heterotopic mirror 6, 25, 28,
        33
   and identity 5, 18, 51, 81, 86, 91
   imagined 97
   and information flows/technologies
        86–9, 90, 110
   non-places 35–6
   of the other 141, 142
   and representation 19, 33, 35, 116, 160
   and space 1–2, 5, 11, 17–20, 22, 24,
        26, 35, 51, 157
   see also body/embodiment, and place;
        Bolter and Grusin, on non-places;
        Casey, fate of place; heterotopia/
        heterotopic, non-places as

     heterotopias; image, the place of
     the; network/ed, place
power
   of differentiation 12, 18–19
   disciplinary 104, 131
   dynamics/forces/forms 9–10, 12, 32,
        88, 100, 120, 122, 133(n)
   of figurations 12, 66
   -geometries 20
   pervasivity of 114
   relations 20, 85, 87, 91, 93, 117, 121,
        133
   to see/visual 12, 60, 63, 80, 100, 103,
        105
   and topography 32, 113
   see also bio, power; difference,
        power-related; Foucault, biopower;
        Foucault, power relations in space;
        grid, of hyper-geo-mapping-
        power; Haraway, technoscience,
        technobiopower; network/ed,
        control/power; system, of power

quantum physics 2, 37, 59(n), 68
   see also entanglement, in quantum
        physics

real/reality
   anti-realism 61, 63
   augmented reality (AR) 105
   in constrained constructivism 63–4, 69
   critical approach to 129–30
   figural 66
   and heterotopias 28
   and language 57
   life 21, 44, 158
   material/realm of things/substantial/
        world 30, 34, 41, 56, 59, 102, 109,
        149, 157
      and symbolic 41
   and mediation 7, 35, 101, 158
   in new technologies discourses 33–5,
        101
   real space 28, 38–9
      vs. cyberspace 24–5
      vs. digital space 8
   res extensa effect 108
   Samantha in *Her* 45–6

and virtual 35, 37, 88, 97, 98, 149, 151(n)
   real virtuality 88
   see also Barad, agential realism; Barad, nonrelativist realist position; representation/al, and reality
recombinant
   art 14
   bacteria 140
   computing 149(n)
   perspective 87–9, 91, 161
   science 126
   technospaces 129
   theatre 126
reflection
   mirror 13, 25, 28, 68, 102, 105
   representational 11, 66–8, 70, 102, 127
      vs. diffraction 11, 60, 67–8, 70
relativism
   as partiality 57
   in standpoint epistemology 55(n)
   and universalism 11, 55, 57
   see also Barad, nonrelativist realist position
representation/al
   and affect 150
   approach/framework/logic/method/paradigm 4, 9, 59, 111, 115, 149, 157, 158, 161
   bias 106
   classical/traditional 4, 38, 68, 114, 127, 151, 157
   and consciousness 49
   consistent 11, 64–5, 69
   of disease 143
   imagination/imaginary 6, 8, 25, 103, 107, 134, 157
   metaphysics/theology 66, 67, 69
   in non-representational theory 1, 11, 61–3
   performative/performed 1, 2, 3, 5, 9, 12, 77–8, 106, 109–10, 116, 119, 148, 150, 158, 160
      and performance art 125–6
      'Opening Conclusions: Performing Represent-actions' 157–62
   politics 63, 66, 79, 104
      and ethics 79

and reality 6, 12, 16, 25, 27–9, 54, 55, 59–61, 64, 69, 70, 78, 100, 101–2, 105–7, 148–9, 150, 157, 160
   realism 24, 101
'Reconceiving Representation' 51–83
relocation of 159–60
and remediation 7, 34–5
represent-actions 157, 160
representability 24, 61, 94, 158
representationalism 1, 11, 29, 38, 59–60, 67, 68, 158–9
   critique of 3, 5, 11, 59–60, 63, 71, 158–9
   in science/technoscience 15, 54, 56, 58
rhetoric 71
and/of space/spatialisation 1–2, 5–9, 21, 23–6, 28, 29–31, 33–4, 35, 38–9, 86, 88, 100, 102–3, 107, 109, 110, 114, 116, 150, 152, 157–60, 162
   'Space and Representation' 17–49
   and time 7, 78, 86, 152
sustainable 100
techniques/technologies 28, 108
and/in/of technospaces 1, 116, 132, 140, 158, 162
totalising 6–7, 11, 33
viable 11, 64, 67
see also Barad, on representation; cartography, representational; computer, and representationalism; diffraction, diffractive representations; displacement, in representation; distance, representational/visual; epistemology, representational; field, representational/visual; generativity, of representation and visual practices; Haraway, representation; Jameson, representational tropes; maps, representational; place, and representation; reflection, representational; scale, as representational practice; site/s, of representation; system, -relative

representation; truth/s, and
representation
responsibility 5, 57, 63, 69, 79–80, 101,
128, 145, 151, 156, 161–2
responsible reflexivity 55–6
see also knowledge, and responsibility
Rich, Adrienne 2, 10
'Notes Towards a Politics of Location'
51–4
see also location, politics of, feminist

Sandoval, Chela
Methodology of the Oppressed 120
scale
as representational practice 90, 121,
160
nano- 94, 121–2
schizo-
schiz- vs. sys- 149, 154
schizophrenia in space 7, 29
and linear perspective 29–30(n)
and psychasthenia 30
see also cartography, schyzoanalitic;
Guattari, schyzoanalitic
Schwartz, Danielle
affective position of the artist 75, 82
Israeli-Palestinian conflict 13, 73–5,
77, 80
Mirror Image 13, 73–83
see also diffraction, in *Mirror Image*;
displacement, mirror in *Mirror
Image*
science/s
citizen 115, 140
end of 108
hard 14, 103
practice 26, 56, 121, 126, 129
spatial 104, 106–7
technological consideration 10, 54, 59
and truth 10, 27, 54–5, 121
see also art, -science; bio, sciences;
creative, science; feminism/
feminist, science (and technology)
studies; Harding, *The Science
Question in Feminism*; knowledge,
scientific; laboratory, scientific;
objectivity, scientific notion of;
optics, sciences; recombinant,

science; representation/al,
representationalism, in science/
technoscience; technoscience
screen
computer (in *Her*) 38–9, 45
detection 2, 71
and diffraction 15, 68
as interface 37, 41, 133
self-replicating automatons (SRA)
in *Teknolust* 8, 43–4
sex/sexual
and the body 52, 121
economies in Biemann's work 93,
95–6, 99
education 122, 123
relationship in *Her* 46–8
Samantha's sexualisation 8
sexism 32, 96, 124(n)
social structuration of sexual difference
91
in subRosa's works 122–3, 126, 128
see also global, networks/flows/routes,
sex trade and industry
signs
different order of 29
different practice of 6, 29
and things 8, 20, 64, 76, 108–9, 111,
150, 157, 158
see also system, of signification/signs/
representation
simulacrum 28, 107
site/s
Casey's definition 18–19
composition of 90
and countersite in Foucault 5–6
of representation 31
sight-seeing vs. site-seeing 104
site-specificity 112
see also situational, site-u-ational
situated knowledge/s 2, 5, 11–12, 53,
56–60, 63, 67, 80, 128, 156, 159
see also diffraction, and situated
knowledges; epistemology,
situated; Haraway, situated
knowledges
situational 6, 9, 19, 31, 63, 89, 112, 118,
129
site-u-ational 4, 126

see also cartography, situational
Situationism 112, 114
   mannerist 115
   revival in locative media 115
   psychogeography 114
sociotechnical
   agents 21
   assemblages 58, 158
   circulation 12, 66
   configurations 105
   environments 1, 6, 61
   formation 89
   mediations 20, 36, 37, 89, 150
   perspective 38
   see also feminism/feminist,
      sociotechnical perspective;
      network/ed, sociotechnical
space/spatial
   absolute/relational 5, 6, 17–19, 25, 53,
      57, 85, 88–9, 157
   abstract-symbolic/material 32, 33,
      113, 120
   becoming space 63
   code/spaces 111
   differences/heterogeneity/multiplicity
      4, 6, 23, 32–3, 36, 85, 120
   digital 2, 8, 25(n), 85, 86, 120
   energy-space imaging 4
   of femininity 91, 96–8
   first and second order space 26–7
   hegemonic/inclusive/privileged 10,
      52, 54
   heterotopic 6, 7, 23
   internal/external 10, 29, 133
   metaphors 17, 26, 27(n), 35–6, 48–9,
      53, 85
   of transparency and opacity 7, 34
   objective/subjective 24, 157
   openness/dynamism 23, 25–6, 116
   politics/tactics 1, 23, 32–3, 51, 53–4,
      120, 121, 129
   practices 7, 35–6, 116, 157
   sociospatial 14, 85, 87, 90, 121, 134,
      135, 137, 138
   spatial turn of media studies 110(n)
   and time/spatio-temporal 4, 5, 6–7,
      19–20, 21, 25–6, 27(n), 29, 32–3,
      35, 67, 76, 78, 83, 85–6, 88, 106,

107, 116, 129, 137, 152, 155, 159,
   160
topology 20–22
unified/unifying 7, 32, 33, 53, 108
urban 29–30, 98, 140
virtual/ity 25, 35, 116
see also art, spatial imagination;
   control, spatial; cyber, space;
   Deleuze, and Guattari, striated
   and smooth space; distance,
   spatial; feminism/feminist,
   spatialisation of; field, spatial;
   Foucault, 'Of Other Spaces';
   Foucault, panoptical model;
   Foucault, power relations in space;
   Foucault, spatialisation of medical
   knowledge; Foucault, temporality
   of space; generativity, spatial;
   geography, as not-merely spatial
   discipline; Graham, perspectives
   on space and information
   technologies; Harvey, time-
   space compression; imagination,
   spatial; information, informational
   space; Jameson, theory of
   postmodern space; knowledge,
   and space; maps, and territories/
   spaces; Massey, spatial theory;
   measurement, spatial; network/
   ed, social/society, sociospatial;
   network/ed, space; observation,
   space of; optics, spatial;
   performative/performativity, space;
   place, and space; real/reality, real
   space; representation/al, and/
   of space/spatialisation; schizo-,
   schizophrenia in space; science/s,
   spatial; technospaces; technology/
   ies, geospatial; vision/visuality, and
   space
subRosa 4, 122–30
   Can You See Us Now? 123
   Cell Track 128
   Epidermic! DIY Cell Lab 126
   Knowing Bodies 126
   mimicry 127–8
   performances 125–8
   Refugia 14, 129

Wilding, Faith 122
see also sex/sexual, in subRosa's works
surveillance
    aerial 117
    art 106, 116–17
    at the border 10, 117, 120
    medicine 131, 133
    sousveillance 106
system
    body as 133–4
    complex 8, 89, 94
    consciousness 49
    distributed 41, 46, 72
    measuring 22, 90
    openness/closure 16, 14, 64, 71, 92,
        132, 139, 150, 161
    of power 105
    -relative representation 61
    paradigm/model/theory 148, 151(n)
    of signification/signs/representation
        21, 48, 92, 94, 151, 157
    of truth 102
    visual 70, 79, 106
    0/1 26
    see also art, and system theory;
        artificial, Intelligent Operating
        System; autopoiesis, autopoietic
        systems; ecosystem/ic;
        environment, system-environment;
        global, positioning, systems;
        information, geolocalised
        information, Geographical
        Information Systems; schizo-,
        schiz- vs. sys

technology/ies
    background/environment 42, 88, 148
    digital 34–6, 90, 108, 111, 123, 134
    forces 5, 86–7
    and gender 98
    geospatial 9
    instrumental view of 7, 21, 36, 54,
        59(n), 149, 161
    locative 9, 111–13, 115–16, 117, 121
    machines 32(n), 132
    of production or reproduction 29
    mobile 139–40
    and nature 72–3, 88, 99

performative 10, 118
reproductive/assisted reproductive
    (A.R.T.), 14, 133
smart 127
and society 6, 7, 21, 29, 36, 85, 88,
    116, 136, 149, 161
technological methodology 54
technological relationship 45
of travel 93
see also aesthetics, and the
    technological; bio, biological/
    technological; bio, technologies;
    control, technologies of; feminism/
    feminist, science (and technology)
    studies; Graham, perspectives on
    space and information technologies;
    information, technologies/
    communication technologies;
    locative media, technical definition;
    maps, performative mapping
    practices, mapping technologies;
    media, visual and visual
    technologies; mobility, migratory/
    transnational and ICTs; place, and
    information flows/technologies;
    real/reality, in new technologies
    discourses; representation/al,
    techniques/technologies; science/s,
    technological consideration; vision-
    visual/ity, devices/instruments/
    technologies
technoscience 15, 66, 68
    postcolonial 72
    practices of 13–14, 56–7, 121, 126–9,
        145, 147, 150
    see also diffraction, 'Diffracting
        Technoscience'; Haraway,
        technoscience; representation/al,
        representationalism, in science/
        technoscience
technosociality, notion of 88
    see also sociotechnical
technospaces 1–2, 6–9, 11–16, 34, 36, 38,
    58, 86–92, 107–9, 111, 113–16,
    120, 121–2, 129, 131–2, 139, 150,
    154–5, 158–9, 161–2
    'Of Other Technospaces' 29–36
    'Technospaces in the Middle' 85–92

see also aesthetics, of technospaces;
connection/s, in technospaces;
cyber, space, cyborg bodies
in technospaces; heterotopias,
technospaces as; location, in
technospaces; mediation, in
technospaces; mobility, in
technospaces; recombinant,
technospaces; representation/
al, and/in/of technospaces;
topological/topology, technospaces
Thrift, Nigel
  a-whereness 88
  non-representational theory 11, 62–3
time
  anything-anywhere-anytime dream 87
  dromocentric epoch 19
  event 5, 9, 19, 25, 62–3, 71, 87, 104,
    125, 153
  of figurations 152
  multiple/plural/specific 6, 65, 116,
    129, 144
  in real time 135, 140
  see also Foucault, temporality of space;
    Harvey, time-space compression;
    heterotopias/heterotopic, and
    heterochronies; Massey, spatial
    theory, coevalness; representation/
    al, and/of space/spatialisation,
    and time; space, and time/spatio-
    temporal
topological/topology 21–2, 30, 85, 109,
    149(n), 160
  approach 36–7, 90
  culture 20–21
  interface 36
  manifold geometry 21
  technospaces 6
  see also space, topology
transnational 13
  actors 93, 96
  alliances 95
  economies and ICTs 9, 92, 93, 98–9
  ideology 72
  perspective 123
  see also borders/boundaries, national/
    transnational; mobility, migratory/

transnational and ICTs; video
  essay, and the transnational
truth/s
  of disease 131
  partial/incomplete/relative/multiple
    13, 41, 55, 65, 72, 73, 78, 81, 92,
    94, 121
  and representation 6, 29, 31, 70, 92,
    94, 102, 103, 107
  see also image, truth of the; science/s,
    and truth; system, of truth
Turing, Alan
  'Computing Machinery and
    Intelligence' 40
  Turing test 8, 40–41
  gender reading of 40(n)
  see also Hayles, interpretation of the
    Turing test

ubiquitous
  commons 134
  computing/information 9, 21, 29, 36,
    85, 87, 89, 105, 111, 134, 148,
    149(n)
  interfaces 20–21, 158
  maps 114
  networks 89, 105–6

Velásquez, Diego
  Las Meninas 38
video essay
  and diffraction 94
  vs. traditional documentary 94
  and the transnational 9, 92–3, 97–8,
    123
  see also Biemann, video essay
virtual/ity 7, 15, 19, 21, 25, 28, 35, 37, 43,
    44, 88, 89, 92, 93, 97, 98, 108, 116,
    117(n), 119, 121, 122, 139, 148,
    149, 150–51, 156, 158
  see also affect, and the virtual;
    Haraway, virtual; real/reality, and
    virtual; space, virtual/ity
vision-visual/ity 9, 12, 14, 31, 60, 62, 63,
    67, 69, 78–82, 100, 101–8, 131,
    149, 152–3, 157–8, 159, 161
  articulation 13, 73, 142, 143, 158

comprehensive/dominant/unifying/
    totalising 31–2, 100, 101–3, 104,
    107, 157
culture/s 65, 71, 104
devices/instruments/technologies 8,
    60, 79, 81, 96, 98, 99, 101, 103,
    104, 106, 108, 113
distance/ing 34, 61, 101, 102, 156, 160
ecology 100
essentialisation of the feminine 125
frog's 64
metaphysics of 62
practices 5, 12, 60, 67, 79, 104, 105,
    130, 142
as re-vision 2, 10, 13, 52–3, 63, 78,
    152
and space 7, 30–31, 34, 101–4, 106,
    107–9, 120
studies 70, 103
viewers'/visual perspectives 9, 12, 25,
    28, 34, 45, 52, 60, 63, 64, 70, 72–3,
    76, 77, 78, 81–3, 92–4, 96, 100,
    101–3, 106, 109, 142–3, 153
visibility 15, 61, 70–72, 75, 80, 87,
    100, 101, 103, 117, 120, 131,
    133–5, 142, 156
visualisation/s 4, 18, 60, 67, 70, 93,
    94, 96, 101, 107, 111(n), 124, 133,
    137

see also body/embodiment, of
    vision; difference, visual/visual
    construction of; diffraction,
    in media and visual studies;
    distance, representational/visual;
    field, representational/visual;
    generativity, of representation and
    visual practices; Haraway, theory
    of vision; media, visual and visual
    technologies; mediation, of vision;
    power, to see/visual; subRosa, *Can
    You See Us Now?*; system, visual;
    video essay

waves
    wave-like pattern 68, 145
    wave-particle duality 2, 4, 68, 159
Web 2.0 25, 85
Women on Waves (WoW)
    A-Portable 123
    Misoprostol campaign 124–5
    see also medicine, medical abortion

Made in the USA
Las Vegas, NV
10 October 2021